Dolphin Diaries

Dolphin Diaries

My 25 Years with Spotted Dolphins in the Bahamas

Dr. Denise L. Herzing

ST. MARTIN'S GRIFFIN

NEW YORK

 For information, address St. Martin's Press, 175 Fifth Avenue, New York, N.Y. 10010.

www.stmartins.com

Design by Kathryn Parise

The Library of Congress has cataloged the hardcover edition as follows:

Herzing, Denise L.
Dolphin diaries : my 25 years with spotted dolphins in the Bahamas / Denise L. Herzing.—1st ed.
p. cm.
Includes bibliographical references and index.
ISBN 978-0-312-60896-5
1. Atlantic spotted dolphin—Behavior—Bahamas. 2. Atlantic spotted dolphin—Research—Bahamas. 3. Animal communication. 4. Human-animal communication. 5. Herzing, Denise L. I. Title.
QL737.C432H464 2011
599.53—dc22

2011005995

ISBN 978-1-250-00691-2 (trade paperback)

First St. Martin's Griffin Edition: July 2012

10 9 8 7 6 5 4 3 2 1

To my teachers Rosemole,
Little Gash, and Romeo

Contents

Part 2

THE MIDDLE YEARS

Observation—1992–96

Cracking the Code: Detection and Deciphering

95

Part 3

THE LATER YEARS

Insight—1997–2008

Two-Way Communication in the Wild: Is It Possible?

173

Acknowledgments

I am grateful for a 2008 fellowship from the John Simon Guggenheim Memorial Foundation, which made possible the writing of this book. Special thanks to my hosts Erika and Triandophyllos at Cheledonia Villias in Oia, Santorini, Greece, for their support during my stay. I couldn't ask for a more special or inspirational place to write.

My thanks to past board members and friends, including Anne Earhart, Diane Ross, Linda Castell, Chris Traughber, Ruth Petzold, Ivi Kimmel, Mac Hawley, Judith Newby, Peyton Lee, Lisa Fast, Richard Reitman, William O'Donnell, Lynda Green, Ginny Lu Woods, and many others.

I am also indebted to my fellow scientists for their consummate creative and engaging work, including Dr. Kenneth R. Pelletier, Dr. Thomas White, Dr. Adam Pack, Dr. Fabienne Delfour, Dr. Lori Marino, Dr. Christine Johnson, and Sir David Attenborough.

Thanks to the many foundations supporting my work over the years, including the Annenberg Foundation; Geraldine R. Dodge Foundation; Henry Foundation; Hawley Foundation; Marisla Foundation; Martin Foundation; Munson Foundation; Pacific Life Foundation; Pegasus Foundation; Plum Foundation; Seebee Foundation; Shaklee; Donald Slavik Family Foundation; Kenneth A. Scott Charitable Trust, a Keybank Trust; Offield Foundation; MAH Foundation; S.E. Printing; and all the loyal members of the Wild Dolphin Project. For early years of support my thanks go to the Cetacean Society International; Kabana; Shaklee the Whale and Dolphin Conservation Society; Oceanic Society Expeditions; and the American Cetacean Society.

Captains, crew, and staff all kept the fieldwork intact over the years, especially Captain Dan Sammis, Captain Will Engelby, Captain Peter Roberts, and assistants extraordinaire Nicole Matlack, Kelly Moewe, Cindy Rogers, and Michelle Green.

Thanks to my editor, Daniela Rapp, and my agent, Wendy Strothman, for leading me through the quagmire of publishing. Daniela's perseverance through the years of fieldwork is especially to be commended.

A special thank you to some of my closest friends and fellow adventurers on this journey, including Chris, Linda, Ken, Diane, Anne, Pat, Lynda, and Adam. You have been my rudder to steady me on my course.

Most of all, my heartfelt thanks to Rosemole, Little Gash, Romeo, and all the spotted dolphins in the Bahamas for sharing their lives. Like other field biologists, I am only a translator, but I am happy to have been given the job.

Foreword

It is my good fortune to have known the author of this marvelous, inspiring book since our days as hopeful undergraduates. Like so many of us back then, Denise had dreams of observing wild dolphins close up, of becoming the Jane Goodall of cetacean studies. The difference with Denise is that she went out and made it happen. And that same relentless determination, which moved her past every obstacle—from funders to sea lice, from engine failures to hurricanes—has led her to become a world authority on the Bahamian spotted dolphins and founder of the Wild Dolphin Project, and has now produced this delightful book for the rest of us.

Denise in the field is the living definition of "intrepid." Tireless and undaunted, her compact, powerful body driving that underwater camera after her fleet and energetic subjects, she always made it look easy. And despite deploying that bulky technology, from which we have learned so much, she could somehow adopt the fluidity and grace of the dolphins she observed, herself almost seeming a natural creature of the sea. Her efforts have advanced our knowledge of a taxon notoriously difficult to study in the wild, capturing in her video and audio recordings the details of the daily lives and long-term histories of these amazing animals. And while her work is recognized the world over for its careful precision, comprehensive breadth, and impeccable integrity, its accessibility in this informative, celebratory volume is every bit in the Dr. Herzing style.

Denise's solution to the impossible problem of how to fund such prohibitively expensive research—boats and fuel and docking fees, crews and cameras and computers—was to open it up to the rest of the world, to allow anyone with the dream of swimming near wild dolphins to come along for the ride. A given expedition might include movie stars, inquisitive students, documentary filmmakers, and just plain folks with a lifelong desire to get up close to a dolphin. But this would never have been the success that it proved to be, and engendered the invaluable research reported here, if not for Denise's uncanny ability to get along with them all, to make each person feel equally important and truly welcome on board.

Also lucky for us, Denise has a remarkable knack for attracting good people, indeed the best! Upbeat, indefatigable boat crews that somehow avoid the pettiness that could so easily arise in prolonged confinement to close quarters, a board of directors with far-reaching influence and resources that nonetheless still manages to keep the animals' interests at heart, and a long list of avid collaborators who recognize not only the great bounty of the clear waters and human-friendly dolphins of the Bahamas, but the qualities in Denise that have made her such a productive and coveted cohort. She strikes the perfect balance between rigorous scientist, sympathetic naturalist, and imaginative visionary. She puts in long hours, not just in the water, but at the computer, archiving and analyzing, keeping a decades-long database under control. She has worked closely with many of the luminaries in the field, studying behavior, acoustics, and even real-world cognition. Her "Phase II" work, attempting to engage in real two-way communication with the animals, is a breakout enterprise, and reveals the true insights she has gained into her subjects. It can be easy to forget the tremendous effort that had to be invested to produce a summary work like *Dolphin Diaries,* but the quality of that effort is apparent on every page.

In this book you will get to know a host of delphinid characters—Little Gash, Stubby, Romeo, Rosemole, and others—in whose fascinating company Denise has spent the last quarter of a century. You'll learn of the stages they go through as they accumulate the knowledge, along with the spots, that mark them as accomplished adults. Denise was there for you, for all of us, documenting their life sagas, watching as generations grew to maturity and then had infants of their own, yahoo-juveniles maturing into significant players in the complex politics of adulthood. And you'll gain something of the feel of what it is to be out on the water, out of sight of land for the duration, where sea and sky become the whole world, in more shades of blue than you ever knew could be.

I have had the unparalleled good fortune of accompanying Denise on a number of research excursions over the years. And it would be no exaggeration to say that these trips have changed my life. The memories I have of them are among the most vivid and moving that I can recall. The mesmerizing ocean! The crusting tang of the salted spray as you pound along searching for fins. Or while anchored and staring at the kaleidoscopic surface that never, never stops moving—a life lesson in carrying on. And open around you, in all directions, the subtle and magnificent skyscape: sweet pale dawns, the glory rays of sunset through the clouds, the stars so thick you can actually sense how far away they are. And a moon that is no disc but a luminous globe so fat it seems it surely must fall into its reflection in the sea. It may sound trite, but it's no less true, that out on the foredeck, late at night, in its silvery light, you find yourself.

There is simply nothing so thrilling or humbling as that moment when you realize that the dolphins have chosen to spend some time with you. You think you're the ones in charge, until you suddenly see that they've taken time from their rich, engaging, demanding lives, to check out what fun can be had with the peculiar humans. How they indulge us, we awkward terrestrials, struggling to hold our breath, pushing our aching muscles to keep up with their effortless grace. It's exhilarating—the giddy thrill of being an animate toy, an object of their curiosity. There's just nothing better than a wild dolphin challenging you to maintain eye contact while it speeds in circles around you, spinning you like a top, until you finally fail to keep up and it breaks away, wiggling in playful triumph. Or the precious honor of being allowed, however briefly, to swim along with a cadre of stately elders. Or the stunning realization that, despite all that has been learned and passed on by experts like Denise, the depth of the remaining mysteries is still, like the ocean itself. . . .

Those of you familiar with Denise's research will have much to savor in this entertaining and enlightening account of her work. And those of you new to her research have a great treasure of discovery in store. But for all that you may learn, perhaps the most important thing that comes across in this book is Denise's respect for her subjects and for the natural world as a whole. The motto of the Wild Dolphin Project is "In their world, on their terms." That says it all.

Christine M. Johnson, Ph.D.
Department of Cognitive Science,
University of California, San Diego
Scientific Advisor on the Wild Dolphin Project

Introduction

In wildness is the preservation of the world.
—Henry David Thoreau,
Walking, 1862

I have been privileged to work with some of the most intelligent animals on the planet—dolphins. More than two decades ago I came to this remote location in the northern Bahamas with hopes of finding a long-term research site to observe dolphins underwater and eventually attempt interspecies communication with a wild pod of dolphins. As a behavioral and marine biologist, I have observed three generations of dolphin families. This book is about that resident community of Atlantic spotted dolphins, *Stenella frontalis*. I have resided in their world for the last twenty-five summers, tracking the lives, deaths, and births of more than two hundred individual spotted dolphins and two hundred bottlenose dolphins, watching them grow up, have fights, develop friendships, and take on the responsibilities within their wild dolphin society.

My work is divided into three time frames based on my own process of understanding and observations. The early years, from 1985 to 1991, I spent figuring out how to observe and interact with the dolphins underwater. I discovered that through a patient, perseverant, and respectful process the dolphins allowed extensive observation of their behavior and social interaction. They let us into their lives and we became privileged observers of their courtship, mating, and childhood play for two decades. Things took their own pace, drifting slowly into the knowledge of another species through observation, interaction, and exploration of their world. Within my work was always an awareness of the dichotomy between leaving wild dolphins alone and interacting with them, a balance I tried to strike while exploring this unique opportunity.

During the middle years, from 1992 to 1996, the dolphins showed us extraordinary behavior and patterns. I began to see complex underwater behavior on a regular basis and to understand the processes by which the dolphins developed these behaviors. In opposition to the fully actuated, adult, ritualized, and predictable behaviors were the "developing juvenile behavior," which contained similar, but uncoordinated body movements and vocalizations, not yet ritualized but varied with intensities of movements and sounds.

During the later years, from 1997 to 2008, I watched many individual dolphins grow up and produce a third generation. Observing and interacting with many of the now grandmothers and elder males of the society was like growing up with an aquatic family. I initiated some advanced projects with high-frequency sound recording equipment, genetics, and cognitive studies, including social learning and teaching. In 1997 I also initiated my Phase II work, developing an underwater keyboard and interactive protocol for two-way communication between humans and dolphins, which led us through four fascinating years of interspecies communication experiments. Then, in 2004 and 2005, the dolphin community was ripped apart and reshuffled from the impacts of Hurricanes Frances, Jeanne, and Wilma.

It is my goal to illuminate the lives of these individual dolphins to you through their triumphs, their failures, their development, and interaction with their environment. These stories will, of course, be through my eyes, which, although trained, will still be limited to a human perspective. Yet we need to bear witness to the truths of the natural world as best as we can. The impact and passage of time only deepens and enriches the trust and quality of sharing during repeated encounters with the dolphins, a process that anthropologists know well. Of course we directly and inadvertently affect the process of our own research. In my case I was in the water, observing intimate details of the dolphins' life, and interacting with these individuals as another species. I describe the process of establishing my research project, getting to know the dolphins, their personalities, and their patterns of behavior. I have tried to share my thoughts and feelings as well as my own triumphs and failures since they were vital learning experiences.

Now their community has shifted and changed, and although it may once again reach some equilibrium after nature's fury, it will undoubtedly be different. How would we know how life has changed without those observations from the last two decades? We simply would not. This is the real value of long-term research.

Dolphins are like icebergs—what you see on the surface is only a small part of the activity underneath. There are stories of the individual dolphins that make up this group. There are stories of their families and their encounters with sharks. There are stories of my encounters with sharks. There are stories of first arriving in the Bahamas and the years it took to win the dolphins' trust. There are stories of data gathering, storm chasing, and people dynamics. There are also stories of juvenile dolphins that grew up in front of my eyes. Some had calves of their own and went on to become productive members of their society. Others struggled and died in the wild. Over the years we documented their range, their movement, and the habitats in which the dolphins spend time. I have seen dolphins hunting on the shallow sandbanks during the day and off the deep edge of the sandbank where fish and squid amass at night. I have watched the fascinating and complex relationship between the spotted dolphins and resident bottlenose dolphins, which entails foraging, aggression, and interspecies babysitting.

Dolphins live in a complex society with friends and relatives: they eat and hunt, raise young, share responsibilities, avoid predators, and resolve conflicts. They also live in a sensory world we can only imagine, full of different sounds, sights, and tastes, and a world we may never entirely understand. What we can share with dolphins is their family lives, their daily challenges, and their fierce devotion to their offspring and community. It is here where the boundaries between humans and dolphins intersect and it is here where we must look for hope in their cultural preservation, as we hope for our own.

I use the technique of "anthropomorphizing" to describe what the dolphins do in the wild and the closest human analogies possible. In this book I tell the stories and share observations of my dolphin work that are difficult to include in scientific papers.* In some instances I have used the tool of anthropomorphizing and ascribing emotions to the dolphins, as justified in Charles Darwin's *Emotions of Animals* and Marc Bekoff's recent book *The Emotional Lives of Animals.* Both make the argument for emotional continuity in other species and its scientific support. As Donald Griffin said in *The Question of Animal Awareness,* a groundbreaking book from the 1990s, "anthropomorphizing is a tool for us to think about what might be going on."

I also discuss my thoughts about dolphins in captivity and the ethics of the dolphin trade. As Al Gore labeled his film about climate change *An Inconvenient Truth,* captive dolphin issues are also somewhat of an inconvenient truth. It is inconvenient because it challenges both the human assumption of uniqueness and the lucrative financial business of capturing wild dolphins for swim programs and human-assisted therapy. I know of very few dolphin-loving humans who would condone these programs if they knew that their child was swimming with one of its victims. If I could, I would apologize to the dolphins for this behavior by representatives of the human race.

Humans have the most complex brains on the planet, and even beyond the great apes are dolphins, with the second-most evolved brain. Yet dolphins have no hands to manipulate objects or build things; the qualities we often subscribe to advanced intelligence. Can we put our

*For technical and scientific papers see www.wilddolphinproject.org library or the selected resources list at the end of the book. Also, http://home.earthlink.net/~dolphindiaries.

imagination and creativity, our best science and technical advances into action to understand another species? This would be a different challenge than we've ever undertaken before, perhaps not all technical but empathic, scientific but participatory, and interactive instead of invasive. The goal of Phase II, interspecies communication, was to attempt to close the gap, to build a bridge of understanding. Like a migratory species I returned every summer to the shallow sandbank of the Bahamas, as I have for the past twenty-five years, still asking that big question: What are they doing with all that complex brainpower in the vast ocean?

What my work in the field has taught me is that dolphins are intelligent beings with complex lives, relationships, and communication. In the wild we have a chance to observe how their cognitive and communicative skills are put to use in the survival game of the real world. It is the closest we can get to living in an alien culture and to gain insight in the process of engaging another species. I am grateful for eager young graduate students who will carry the torch a while longer. Even then we will have only glimpsed the tip of the iceberg.

It is with great pleasure that I now take you under the water with me, although you'll be able to stay dry and free of jellyfish stings and salty hair. Or maybe you won't. Perhaps you'll feel the force of the winds as they kick up the gin-clear seas. Possibly you'll imagine what it must feel like to have an annoying little remora sliding around your body or a large shark follow you back to the boat. It's up to you. Let me take you there to the Bahamas and underwater with the incredible community of wild dolphins I have known for twenty-five years. It would be my privilege.

PART 1

The Early Years

Exposure—1985–91

First Contact with Dolphins: A Vision of the Future

In 1985 I set out to explore an area in the Bahamas that was known to be home to a group of friendly dolphins. My purpose was to assess whether this field site had the potential to support long-term underwater research. It was clear to me, from both Jane Goodall's and Dian Fossey's primate work in the wild, that if you spent enough time in a peaceful, benign relationship with an intelligent animal group you would come to learn from them, and, perhaps, eventually be incorporated into their society. Long-term field research had been done with primates and elephants, why not dolphins? It seemed possible to follow a few generations of dolphins with a minimum commitment of twenty years.

During these years I learned the challenges of working from a boat in the open ocean, working as a young female scientist in this male-dominated field, and the importance of continuity. I started a field project, using various scientific tools and rules of etiquette I learned directly from this aquatic species. It took five years to gain the trust of this family of dolphins. Only after this initial period of habituation did the dolphins start to show us their natural behavior.

I met Little Gash, Paint, and Romeo, individuals I would follow during twenty-five years of fieldwork. And it was only the beginning.

~ 1 ~

First Contact with Dolphins: Establishing the Relationship

If you have the desire for knowledge and the power to give it physical expression, go out and explore.

—Apsley Cherry-Garrard,
The Worst Journey in the World

Pre-1985—Beginnings

If you could ask a dolphin one question, what would it be? And what do you think a dolphin might ask you? These questions occurred to me when I first met a spotted dolphin in the wild early one humid summer morning in 1985. I swam slowly away from the boat that was anchored in the gin-clear waters of a shallow sandbank in the Bahamas. It was calm and peaceful out in the middle of the warm amniotic salt water with no land in sight. Two dolphins approached and swam around me, looking directly into my eyes. There is nothing comparable to making eye contact with a wild creature; it is like a sharp splash of ice-cold water on the face. I sensed a keen and mutually exploratory awareness; I sensed another "being" behind those eyes. Ten years later, after experiencing strong currents and large sharks, I would have a different type of respect for the ocean, one that probably wouldn't allow me to swim out so far alone in these waters with such a calmness. But this experience was different; it was my first encounter with a wild dolphin.

In all my years of work with marine mammals, nothing had prepared me for this. I found myself deeply regretting never taking an anthropology class. What is it like to meet and experience a new culture for the first time, a nonhuman culture? What do you do if they are curious and want to observe you? I was a biologist, a cetologist, who studied whales and dolphins. What brought me to the Bahamas was curiosity about the lives of wild dolphins, but the experiential part was not something I was trained for as a scientist. But this experience seemed perfectly natural. My ancestors evolved with plants, animals, and the Earth itself. Well before that, dolphins, as early mammals, returned to the ocean from their land ancestors twenty-five million years ago. This world of a highly evolved mammal was a window into the dolphins' unique aquatic world, not separate and estranged as the land and sea seem in the open ocean, but intertwined like a shoreline: mutually curious species carefully considering each other.

In the field of animal behavior the philosophy of "to know a goose, become a goose" was first formulated by Konrad Lorenz, considered the father of modern animal behavior. This level of participation has been productive for the study of many social species, including chimpanzees by Jane Goodall, mountain gorillas by Dian Fossey, and African elephants by Cynthia Moss. These pioneering women researchers provided solid examples of a productive way of illuminating the lives of wild animal societies. That is the approach and methodology I decided to use for studying free-ranging dolphins.

For years scientists had attempted to teach nonhuman animal species, including dolphins, the English language, without first learning about the dolphins' natural communication system. I was always fascinated by the idea that dolphin minds evolved in the aquatic environment, parallel, but potentially dissimilar to our own. What would that mind be like and how would it express itself? Could we understand their type of consciousness by studying their communication system and cross over that interspecies boundary? Could we really build a bridge? I decided to focus my work first on understanding how dolphins communicated with each other, using sound, vision, touch, and second to use those same natural channels of communication to explore the possibility of interspecies communication between humans and dolphins.

I grew up in the Midwest far away from any ocean, with the exception of the world of Jacques Cousteau that flowed from our living-room TV. I became both fascinated with and committed to the exploration of dolphins, potentially one of the most advanced nonhuman intelligence on the planet. My passion for this work began when I was twelve years old. I entered an essay contest for a scholarship in my hometown in Minnesota. One of the questions was, "What would you do for the world if you could do one thing?" My answer? "I would develop a human-animal translator so that we could understand other minds on the planet." As I continued to grow up and observe the natural world around me I became more and more fascinated by the idea of complex minds like ours evolving in the water. What could dolphins possibly be doing with all that brainpower if they didn't have hands? I knew then my lifework had been chosen. Simple questions often have the most important repercussions. Where did I see myself in five years, what kind of environment did I want to be in, what kind of people did I want to be around? I wanted to work at sea, in the environment where the animals lived. I wanted to be doing research—to be observing and documenting the lives of wild dolphins. And I wanted to be around stimulating people who could use a variety of talents to explore this unknown territory. These simple answers guided my decisions for the next ten years through graduate school and during the formation of my own research project and nonprofit organization, the Wild Dolphin Project, which became the support structure for my research with the Atlantic spotted dolphins.

So by the age of twelve I knew I wanted to study dolphin communication. After taking the advice of a wise college counselor, I left my hometown to venture forth into the world of marine biology, to get into the mud and see if I liked the field. I applied and was accepted to both the

University of Washington, which had an undergraduate program in marine biology, and the University of Miami, renowned for its oceanographic work. But I decided to go to Oregon State University for two reasons. First, I thought Oregon would be a beautiful place to live with its green landscape and healthy lifestyle. Second, Oregon State had a marine mammalogist, Dr. Bruce Mate, and I wanted to study marine mammals. I jumped on boats whenever I could for oceanographic cruises off the Oregon coast or on salmon fishing boats with friends. I loved the smell of the ocean, the smell of seaweed, the roar of the waves. I discovered that I did, indeed, like the mud.

After completing four undergraduate years at the inland campus I spent my last year at the Marine Biological Station in Newport, Oregon. I observed harbor seals and sea lions at salmon hatcheries. I joined Bruce down in Baja for two winters studying gray whales; the first year with Jim Sumich, a Ph.D. student at Bruce's lab doing work on the metabolic rates of gray whales. San Ignacio Lagoon and other lagoons in Baja California, Mexico, are home to friendly gray whales that gave researchers close access and unique opportunities for studying the species. My second winter I helped with radio-tagging and monitoring from land with receivers for previously tagged whales. One day the field team came back to shore and I excitedly told them I had heard Blanco's radio signal, the first whale tagged. After the winter tagging season we quickly placed one of the radio receivers in a lighthouse on Yaquina Head, the Oregon headland where I had previously spent three years counting migrating whales. One night, back in Oregon, I heard Blanco's signal—he had gone by the lighthouse! We hopped in our cars and drove, while Bruce, a skilled pilot, flew, and followed Blanco up the coast. Blanco was eventually tracked all the way up to Unimak Pass, Alaska. All of these studies were of great value to me and provided opportunities and insight, and Bruce was an invaluable mentor and teacher.

Around that time I got wind of a beautiful 134-foot wooden barquentine (a form of square-rigged sailing vessel) named the *Regina Maris* and her six-week student program at sea. Taking advantage of her time in the

Pacific, I jumped onboard *Regina Maris* for a six-week adventure. We spent most of our time in Magdalena Bay, another of the gray whale breeding lagoons in Baja. I met Dr. Kenneth Norris (the father of dolphin studies and a future mentor), Kenneth Balcomb (a renowned killer whale field scientist), and other well-known marine mammal scientists. Our chores were typical of being on a vessel at sea. As students we attended classes during the daytime and did watches four hours on, four hours off, to help run the ship. We took classes in celestial navigation and basic seamanship—we handled the wheel on deck, we hoisted sails, and of course we tied knots. Bolen knots, sheepshank knots, square knots—any kind of knot. We felt nautical! The training of young marine scientists should always involve extensive fieldwork because such experiences are critical for understanding the sea itself. I was developing a deep love for the sea, for boats, and my life was now set on a course through the waves. I loved studying the ocean from the small microscopic creatures to the large marine mammals off the Oregon coast.

Then in 1981, shortly after I had finished my last year of undergraduate research on migrating gray whales, I had a nearly fatal accident. While out in the woods alone I fell off a deck of a house and down a ravine, and was subsequently smashed by the falling lumber and deck, breaking my ribs and severing the hepatic artery in my liver and nearly bleeding to death. After barely making it to my twenty-fifth birthday I realized there is nothing like a near-death experience to make you reprioritize your life. So once again I decided to join the *Regina Maris,* now in the Atlantic. With barely healed broken ribs, but my doctor's approval, I boarded *Regina Maris* in Gloucester, Massachusetts, this time not as a student, but as the " lowly student lab slave"—a dollar a day with room and board for six months. Our trip took us directly to Bermuda and down through the Caribbean for our final destination off the Dominican Republic to study humpback whales on Silver Banks, the remote offshore area dotted with undersurface reefs and rocks where humpback whale mothers spend three months birthing their calves or mating.

There's nothing like being in a large sailboat, hearing the wind and feeling the pounding waves. But it was October and hurricanes were brewing in the tropics to the south of us. As the days progressed I noticed the crew getting nervous as they watched the winds picking up. Sails were lowered and hatches battened down, closing off the incoming water. By the end of that day it was clear we were in the tail end of a hurricane. To make matters worse the bilge pumps, the usually automated pumps that keep water to a minimum in the hulls, stopped working. For twenty-four hours straight, rotating shifts, we hand-pumped from the deck, through rain and thirty-foot waves we pumped and pumped. The crew tried to heave to, setting the sails to hold the boat in place and to keep the boat steady in the wind. As the boat rolled on thirty-foot seas I was one of the few students on deck with my trusty Nikonos waterproof camera. I attached myself to the lifeline on deck and took pictures. It was both majestic and humbling. I felt like a cork, going up and down as the seas swelled and our vessel rose and fell again. After two days hove to we limped into Bermuda. But we had been lucky. Small sailboats were towed in from offshore, gutted, broken, and lifeless. This was life at sea, or death at sea, depending on your luck. I guess it was something I would have to get used to.

When it was finally time to leave Bermuda we headed south through the islands of the Caribbean and stopped at a small uninhabited island called Hogsty Reef. Our task here was to measure and count the amount of tar washed up on this remote island. There was tar on the beach, there was tar in the seaweed—there was tar everywhere. It was 1981 and already pollution was present, even on this remote island. We filled our days documenting and cleaning up the tar. Then one day, as I maneuvered the small inflatable boat toward the island, I saw the students waving frantically. I was heading right for a breaking reef invisible to me with the glaring sun. Although I did little damage to the reef itself, I broke the pin of the outboard engine and felt humiliated, unworthy of my position as a lab slave. That night after crying on cook Erma Colvin's shoulder, Perrin Ross, my understanding and skilled mentor, pulled from under his vest a small can of beer to share. Beer onboard the *Regina Maris* was a treat only dispersed on Saturday night, when the crew made trades and swaps, desperate to secure an extra beer. It was a gesture of understanding I would never forget.

It has taken me quite a while to learn to have patience and be sensitive to mistakes that young students will make on board my own research boat in the Bahamas. They forget to charge a battery, drop a piece of equipment, or leave the camera lens on. I also tell my students to expect problems in fieldwork. There will be weather days, your equipment will fail, you won't have enough funding for certain pieces of equipment, and you will just have bad hair days (usually involving salt). You need to expect it, need to be ready for it, and need to always be as prepared as possible because field time is valuable and expensive.

I'm always reiterating the need for redundancy on my boat. I have two cameras, two underwater housings. Film is cheap, videotapes are cheap, but field time is expensive. Use the time and use it well because it may not be there the next day. In twenty-five years of fieldwork I've only flooded one video housing, which is still a good record. Only once have I managed to copy over a piece of video footage. Sadly the footage was quite critical and when we had reviewed it the previous night I had not forwarded the tape for the next day. Again, after twenty-five years, that's not so bad, but we scientists value our data and our experiences in the field tremendously and any loss of data is a potential issue, such as the photograph of a long-lost animal or an unseen behavior lost to a mistake on the video. But at the end of every field trip, in every field season, if we get back alive and the boat is intact, it's been a good summer.

After working with gray whales both off the Oregon coast and in Baja California, and finishing my degree, I was ready to move on to my graduate work. But there was one thing I wanted to do first; I wanted to travel to gain some experience in a country that didn't speak English and wasn't Western. I knew I wanted to study communication. I knew I needed skill sets perhaps beyond scientific training. Luckily at that time my sister decided to buy my half of our family house. It was a small and modest amount, but the money she sent as a deposit allowed me to stash away half for graduate school and half for a three-month trip to Asia. I departed with a backpack, no credit cards, and an awful lot of luck. I traveled to China, Nepal, and India, learning much about nonverbal communication and the universals of cross-cultural experiences. On the way I stopped in Japan to visit my colleague Masahara Nishiwaki, whom I had met in my sailing days onboard *Regina Maris.* The insight gained from these non-Western human cultural interactions, their similarities and differences, was invaluable. We really need so little when we travel (beyond clothes and shampoo) and we can communicate through body language and laughter quite often when language fails. I arrived safely back in the United States, now ready to give my life to my future graduate work and what I hoped to be my future career studying dolphin communication.

At this point in my life I was familiar with the work of three researchers studying dolphin communication. Louis Herman in Hawaii was famous for his cognitive and experimental work. Diana Reiss at San Francisco State University was studying dolphin communication in captivity. And John Lilly, a controversial scientist (but likely a visionary before his time), was exploring two-way communication. So I struck out to the San Francisco Bay Area. I had eliminated the idea of studying in Hawaii since Herman's work was experimental and not really focused on the communication aspects I was interested in. I arrived in the Bay Area and immediately went down to MarineWorld Africa USA where both John Lilly and Diana Reiss had their labs.

I have always had a clear vision of what my life would look like in the future. One of my visions for recognizing my future dolphin work was to walk into a facility and see a spectrographic analysis machine, which analyzes dolphin sounds. When I entered Dr. John Lilly's laboratory, I found the opposite. The staff seemed unclear and unfocused. Although the work was creative and interesting, I knew it wasn't for me. But after arranging a meeting with Diana I walked into her temporary trailer, saw a spectrograph machine, and immediately knew that this was my place. Although my goal wasn't to work with captive dolphins, I learned the importance of correlating sound with behavior since dolphins are, after all, acoustic animals and I learned about the complexity of communication signals. Most humans communicate in similar ways, but anthropological studies prioritize finding out the "meaning" of the signals. Who is making what signals and what is their relationship? Are they males or females, a mother and brother, or are they unrelated? The way to understand human communication signals is to put them in the context of human societies, networks, and relationships, and I imagined it was the same in an aquatic society. I knew I wanted to take a new approach and I knew that my work would involve a broader perspective than animals as subjects or machines. I wanted to take a look at dolphin communication with dolphins being a cultural animal and a member of a unique, intelligent society. I reasoned that this anthropological approach might be helpful. Little did I know how much the participatory approach would be critical in setting the tone of my work through the years.

Shortly after my San Francisco trip I ventured up to the Seattle area to check out one more person, Jim Nollman. Jim was known as an interspecies guy, he played music to orcas, to turkeys, and probably to other species. When I knocked at his door in Seattle I was greeted by a man who was clearly skeptical of science and scientists. The first question out of his mouth was, "Are you a scientist?" "Well," I stuttered, "I'm hoping to be." It was clear that this was not a scenario for me. Years later, while working in Diana's lab, Jim called to ask for her assistance in the analysis of a complex sequence of sounds, playback and response, with his orcas. Although I was glad to hear that Jim had finally learned to appreciate the scientific analysis that might be needed to truly understand interspecies communication, I thought it ironic. To establish a nonspecies-biased science we need to open our minds to other sciences, other disciplines, and other ways of thinking. And this applies to scientists as well. I had been exposed to a cutting-edge scientist and his work in the mind-body health field when working with Dr. Kenneth Pelletier in San Francisco. During my time assisting Ken in his research studies of the brain and mind, I watched as he walked a tightrope between traditional medicine and alternative medicine, eagerly using them to merge the possibilities. Today, Ken is a leader in the field of bridging the two. I knew this would be my challenge with exploring dolphins in the wild. I, too, wanted to bridge a gap.

But where, I wondered, was a location where I could study wild dolphins underwater and in an accessible environment? While I knew of no one who was working in such conditions, I knew somewhere it existed, but how to find it? The answer, like Jacques Cousteau, flowed into my consciousness through my living-room TV. I happened to see a documentary by filmmaker Hardy Jones on a group of friendly spotted dolphins in the Bahamas. I was intrigued by his underwater footage of these wild dolphins. Could these dolphins be easily and regularly accessible to allow long-term observations? This was a potential long-term field site that might allow an opportunity for observing what dolphins do, underwater, in the wild. The water is warm, clear for viewing, and the dolphins seemed uniquely curious about humans. I thought, "Surely there must be someone out there studying them already!" I called Hardy on the phone and he agreed to let me see some outtakes of his film so I could gain a better sense of the possibilities. So in 1985 I ventured out to the Bahamas for six weeks to ascertain if these dolphins were really as accessible as I suspected. These dolphins were discovered by treasure divers in the 1970s and they subsequently befriended the working treasure hunters anchored on the shallow sandbank looking for lost Spanish galleon ships. Their sand-blowing equipment exposed tasty fish morsels for the dolphins that came around occasionally for snacks.

I wanted to know if one could see natural dolphin behavior on a regular basis. As I reviewed Hardy's footage it was clear to me that the Bahamas was a place where a fragile, terrestrial human could work in the water for extensive periods of time and observe behavior underwater, and where the dolphins were accessible because of their curiosity about humans. And to my amazement no one was out there studying them scientifically!

At this point I was fairly convinced that this community of dolphins was an ideal group for a long-term research project. But I wanted to go see it for myself, to assess the true potential for long-term work. Luckily, an organization I had worked with on gray whale research, Ocean Society Expeditions (OSE), was starting to run ecotourism trips out to this area and they agreed to let me go out as their naturalist for a month and explore the possibilities. But what equipment should I bring to the field? I readied my Nikonos camera for photo-identification work and I acquired, through the eager support of a friend, an underwater video housing in which I placed a video camera and an external microphone, called a hydrophone, to simultaneously record the dolphins' vocalizations and behaviors. If there was one thing I learned from graduate work, it was that dolphins communicated in many sensory modalities and that to record them simultaneously was critical to understanding the context of their communication. It was also important to know the individuals, their relationships, sex, and history to make sense of the greater context of their lives. So my plan was to spend time with this wild group of dolphins, figure out who was female and male, identify them as individuals, record their behavior, their associations, document the sounds they make with various behaviors, and then try to make sense of it all. Not so easy in the end. How does one really prepare for meeting another species in hopes of establishing a long-term relationship to allow such an intimate glimpse? I knew of no road maps except the work of Jane Goodall and others who applied patience, perseverance, and appropriate etiquette over time to build trust to access a nonhuman world. These were the models I would follow to enter the dolphins' world.

My plan was simple. I would get in the water, try to stay detached and log identification marks on an underwater slate so I could start identifying individuals. I would try not to disturb their behavior and keep to myself. But as I soon discovered, the dolphins had other ideas.

1985—First Contact

It was 7:00 A.M., the water slick and calm, as it often is during the hurricane season. Out of the topaz blue haze appeared two dolphins, one large, one small, swimming side by side, scanning me as they moved their heads and sent clicks of sound toward me as they approached. I froze, not in fear, but in awe. In all my years of work with marine mammals, the shock of this first contact with another intelligent animal was new. I felt as though I was experiencing a new culture for the first time—a nonhuman culture.

Today I was in the Bahamas for the first time, meeting the spotted dolphins. Oceanic Society Expeditions (OSE) had finalized their trips and I was here for six weeks as an additional naturalist. My personal goal was to scope out this area for long-term research. OSE had chartered a lovely sailing catamaran, the *Dolphin,* owned by Larry Vertefay. Larry had heard rumors of this group of wild dolphins and decided to reprioritize his own successful business in order to explore them. It was a match made in heaven with his lovely sixty-foot boat and OSE trips.

Although I had worked in the field with gray whales, harbor seals, and humpback whales, I had always collected data from the surface. I had never tried to collect data underwater with dolphins. Luckily my longtime friend, Linda Castell, saw the twinkle in my eye before I left for the field. Linda and I had met at the marine lab in Newport, Oregon, in the late 1970s. She was studying marine microbiology and I was working with gray whales. After becoming friends, she must have recognized my determination to do work in the wild, and Linda, being someone who puts her money where her mouth is, without a second thought wrote a personal check for a thousand dollars for me to buy an underwater video system with hydrophone. Like many artists during the Renaissance, scientists are often supported by interested individual donors, and this has been my story.

On board the *Dolphin* we sailed from West Palm Beach, Florida, to reach the sleepy port of West End, Grand Bahama Island. After clearing customs we headed up to the study site, some forty miles offshore and out of sight of land. I am a bit of a Girl Scout when it comes to being prepared, so I was ready with my video system. It was clear to me from the beginning that I wanted to study dolphins like Jane Goodall had studied chimpanzees. I wanted to be a benign observer and get to know the individuals and society through watching their interactions. I wanted to use photo-identification methods to find unique physical characteristics to track individuals and learn about their communication signals and social rules to fit in as much as a human could in an underwater world. I was determined to work with another intelligent species using participatory science, incorporating them as mutual participants in the process, as opposed to traditional science that would view them as subjects and nonparticipants. There was no exact road map for the work, but I knew I needed to engage both my scientific training and my own knowledge of human culture and interaction in the process. If I spent enough time in a mutual relationship with an intelligent animal society, I might come to learn from them, and perhaps eventually be incorporated into their community. Although an unusually long time to commit to a specific field project, it seemed both necessary and possible to aim for twenty years at this field site to document a few generations of dolphins. But this year I only had six weeks.

This summer I wasn't in charge of our anchor locations, how we operated with the dolphins, or anything else. Larry Vertefay and his crew and the primary OSE naturalist handled most of these details. I focused on collecting identification shots of the dolphins with my underwater camera. I struggled with an underwater slate long enough to know that it was useless with such fast-moving animals, and focused on my underwater video to capture sounds and behavior. Although I had learned some basic communication signals from dolphins in captivity, it was difficult sampling dolphin signals in the field due to their fast swimming and complex behavior.

During the days at anchor the dolphins cautiously came by the boat to check us out. We let them explore us, on their own time and in their own way. We slipped into the water slowly, trying to be nonaggressive and cautious in order to gain their trust. Although I got occasional glimpses of the dolphins chasing and fighting each other, it took another five years to build their trust and see their normal dolphin behavior. At night at anchor under a sky pregnant with stars we listened to the dolphins with our underwater hydrophone. Interspersed with the lapping of waves against our hull we heard a cacophony of sounds complex enough to make us think that the dolphins themselves were engaged in a detailed discussion. We speculated on what was being said and what the dolphins might be doing during such vocal exchanges. Later I came to understand more about these sounds and their complex nature and could predict the behavior associated with these sounds. But this first summer, these sounds could have been anything and I wondered what information the dolphins might share with each other. Do they talk about their day or only express what is happening at the moment? Although I knew that most dolphins made signature whistles (a type of frequency-modulated whistle unique to an individual) for contact calls, echolocation clicks (sonar for orientation and navigation), and burst-pulsed sounds (a clump of clicks in discrete packages used for close proximity social interaction), it took years of observing their underwater behavior to understand the subtleties of these communication signals. I have never regretted my time in the field observing, over and over, the same behaviors. It is in the field where we actually see what is happening. My commitment was to observe the dolphins firsthand to most accurately interpret their behavior.

To study social animals it is critical to follow individuals in the larger group. When I began the work one of my first priorities was to document individuals by their natural marks that were consistent and unique. Photo-identification had been used with chimpanzees, elephants, zebras, and giraffes. And the technique had been recently applied to whales and dolphins in the field. It was also critical in these early years to sex the dolphins because knowing the players, by identification marks and their sex, was essential for understanding dolphin behavior over time. In my first weeks out there I used my underwater Nikonos camera to photograph and sex as many individuals as possible.

Three of the first dolphins I met were Little Gash, Rosemole, and Mugsy. A tight trio, these juvenile females surfed the waves near our boat one day while I was in the water. With a glint of mischief in their eyes, they swam around me whistling and cavorting. One dolphin had a small nick in the lead edge of her dorsal fin, so I named her Little Gash. The female next to her, although young, already had a large black spot, in the shape of a rose on her right side and she became Rosemole. The third female had a bullet-shaped wound on her body, and I named her Mugsy, a seemingly gangster name to me. These three females were inseparable for the summer, always playing and getting in trouble together. It was rare to see one without the others. This trio of tricksters swam circles around me whistling excitedly, they dropped sargassum in front of me to solicit keep-away games and chases, and they watched my awkward human swimming in the water. Although I would never know who their mothers were, since they were already juveniles and not regularly with an older female, these three dolphins would provide the first example of how strong the bond can be among dolphins, and how it changed over time as they matured and became mothers themselves.

Sexing a dolphin in the water was a bit trickier than identifying their marks. As Rosemole swam by me and turned inverted I glimpsed her two mammary slits, which had identified her as a female earlier but that I knew I needed to recheck. The males that chased her had no mammary slits but were visibly aroused with erections as they chased Rosemole to mate. Week after week I caught glimpses of genital areas and matched their sex with their identification photographs and verified it over and over again. I marked my flippers with an *M* and an *F* and shot the video to record an individual as male or female, a handy technique for in-the-water work. It was really the only way to sex the spotted dolphins since females and males are about the same size, with no obvious physical traits to distinguish them. The only sexually dimorphic trait they have is the very white-tipped beak or rostrum, which males develop as they become older. During rut, when the males have peaks in their reproductive cycles, the older males display something that looks like a ventral keel, a bump in their genital area that is actually an increase in testicle size. Male testicles may double in size during this time and ventral keels often make the males look like pregnant females.

Today as I watched a large group of young adult males courting and mating with some female dolphins, suddenly, out of the periphery, two large old adult male spotted dolphins, Sickle and Pyramid, dashed in. Displaying their full ventral keels these two males rapidly and forcefully mated with the females. "Wow," I thought, "were the young males just prepping the females for the old males?" Wouldn't that be interesting? I had suspected, even after a few weeks in the field, that the behavior within these large groups of young adult males were a male-to-male competitive display for one another and the observing females. To this day we still don't know when Atlantic spotted dolphin males become sexually mature. From our later work with genetics we have verified that many of the fathers are indeed older males at least twenty years old (dolphins can live to fifty years or so) in the group. And today these elder dolphins seemed to have the mating edge.

One of my first big lessons in "dolphin etiquette" occurred during my anxious attempt to sex a dolphin. It was midsummer and I was in the water with a couple of different dolphins. As I dove down underneath one of the dolphins to sex it, another dolphin moved in and placed himself in between the dolphin and me, and then he swam off rubbing pectoral flippers with the still unsexed dolphin. I later came to understand that the action of inverted swimming underneath a dolphin can be aggressive or an invitation to have sex, which I had inadvertently been soliciting. This was my first experience with dolphin etiquette and one I did not forget. After this mistake I learned to be more aware of what messages I was sending as a human in the water. After all we were "in their world, on their terms" and that meant learning, and observing, the local customs.

As we passed the sweltering days anchored on the sandbank the dolphins started showing up at regular times, at 7:00 A.M., 11:00 A.M., 4:00 P.M., and sometimes at sunset. We anxiously donned our fins and snorkels whenever they arrived in hopes of a fruitful encounter. Trip after trip I introduced new eco-passengers to the etiquette of the dolphins and our in-water procedures. They helped us gather photographs, stood watch on the bridge to find dolphins, and watched video with us at night as we reviewed the day's adventures. Some trips were during ideal weather, allowing us to anchor for long stretches. During other trips we ran from swells and storms that broke our routine but allowed us glimpses of dolphins in other areas on the sandbank. Occasionally, and sometimes out of boredom, we set sail for a nearby reef for a snorkel, or to try our own human fishing techniques to provide dinner. With a limited amount of fresh water onboard we returned to port crusty and salty. One quick shower at the marina and life seemed normal again.

Already I was seeing the full range of age classes for spotted dolphins. I met Apollo, the calf of Luna, who showed clearly that when born this species has no spots and resemble young bottlenose dolphins with their two tone gray and white coloration. Born with fetal folds, and little knowledge of the ocean like other calves, Apollo stayed close to his mother the first few months of life. Only a quarter the size (about one and a half feet) of the fully grown Luna, Apollo depended on the rich milk provided by her. Apollo, quite precocious and already cavorting with other calves his age, was exposed to the day-to-day activities not only by his mother but also by her many associates. In his nursery group Apollo was already learning to forage and to catch easy prey items, including flounder, on the shallow sandy bottom.

Dolphins create a variety of sounds, including frequency-modulated whistles, clicks, and burst-pulsed vocalizations, using multiple air sacs situated below their blowhole. These sounds are shunted out of a fatty structure in the front of their head, termed the "melon." When their echolocation clicks hit an object, the returning echoes are received through the lower jaw full of fat that conducts the sound to their inner ear, which is basically a mammalian design with a few specializations for hearing high frequency and isolating sounds in the water.

From my graduate work I knew that dolphins had unique whistles analogous to names, thus named signature whistles. Luna had a strong signature whistle and Apollo rapidly learned not only to recognize his mother's whistle, but also to produce his own unique whistle. Signature whistles are used in three contexts: in mother-calf reunions, during babysitting, and in courtship displays. Most of the whistle is audible to a human in the water and it is often correlated with bubbles coming out of the dolphin's blowhole.

There is a variation of a signature whistle, called an excitement vocalization or whistle squawk, which contains components of the individual's whistle as well as a burst-pulsed component. Today Apollo's was out of control and excited, creating his squeaky excitement sounds, and Katy responded accordingly. Katy was four years old, the phase termed "speckled," when the dolphins develop black spots on their ventral side (underneath), newly found independence from their mothers, and responsibilities of babysitting. Excitement vocalizations, used during periods of distress and excitement, attract the attention of the mother or babysitter who quickly join the excited youngster and calm him down with a gentle touch of their pectoral flipper to the calf's body. Today this was Katy's job and she executed it with precision, calming Apollo down quickly and initially without resistance. Apollo continued to squawk and squirm as Katy again struggled to calm this quite rambunctious and out-of-control youngster. Suddenly from the distance Luna zoomed in, swam to her calf, and gently touched him with her pectoral flipper, immediately calming him down. This was the first of many observations of mothers and babysitters and their techniques of aquatic discipline.

My initial goal had been to find a place in the world where I could observe dolphins underwater, but as it turned out Atlantic spotted dolphin (*Stenella frontalis*) were also an ideal species to study because of their changing spot patterns with age. When I first visited the Bahamas, little was known about this species. In fact the taxonomy of spotted dolphins was quite a mess. Now we know that spotted dolphins can have quite different coloration and spotting patterns and still be the same species. For example, Atlantic spotted dolphins living in the Azores in very deep water have few or no spots. Spotting is a camouflage for the light play found in shallow areas like the Bahamas; so the deeper the water, the fewer spots. Originally called *Stenella plagiodon* this species is now identified as *Stenella frontalis*, the Atlantic spotted dolphin, and is found only in the Atlantic Ocean, from New Jersey in the northeast United States down to Brazil and from the shores of North Africa to the mid-African continent. After my first season in the field I began to modify Bill Perrin's spotting categories (the fisheries scientist who identified the age classes and coloration patterns of the pantropical spotted dolphin, *Stenella attenuata*), to better reflect the life history of the Atlantic spotted dolphin.

I watched as Rosemole and Little Gash played and explored together while they learned the art of socializing from older dolphins. I wondered if I would see these juveniles mature and have offspring of their own. I met White Patches, a regular babysitter of both Rosemole and Little Gash, who showed the obvious signs of female sexual maturity, her "mottled" spotting pattern now in full bloom with white spots on her dorsal area in addition to her dark ventral spots. I watched older males, Romeo, a wise old male who not only was a fully grown spotted dolphin with his "fused" or coalesced black and white spot pattern, but a leader in the group, and his friend Big Gash, both court White Patches and attempt to mate with her.

Over and over throughout this first summer I was reminded how important it was to know the sex of each dolphin to interpret behavior. One day I observed a young female, Priscilla, swimming underneath a well-known male adult, Knuckles. Priscilla was in a classic position that mothers and calves display, with the youngster underneath the mother. But today Priscilla was underneath a male adult and if I had not known that this dolphin was a male, I might have mistaken him for a mother with a calf. Today, Knuckles was babysitting the young Priscilla, a less common but nevertheless occasional duty of males. Assuming the sex of animals has been a problem in other underwater studies. In early years of humpback whale work, researchers observed a second adult with a mother humpback whale and assumed that the second adult whale was an attending female. It made a nice, neat human story, but it didn't represent the natural world. Until researchers in the water began to verify the real sex of individual humpback whales, they assumed this to be true. The second adult whale was later verified as a male whale that typically attends females for a mating opportunity. How different that story became when we knew the facts. It was a continuing reminder to make sure that we verified the identification of individual male and female dolphins during behavior events.

Twenty-five years later my graduate students read my old notes from this early time in the field and laugh at the changing ideas and data sheets. Yet no one had ever worked underwater in this scenario, so the designing of forms and information to collect were new, and I evolved them as needed over the years. And that included resexing individual dolphins over and over again for accuracy.

As my six weeks flew and the end of my summer neared, I started noticing very pregnant females and suspected that they would give birth in the fall. Other females were looking only slightly girthy, likely springtime births. By then I had also met Stubby, an individual with a chopped-off dorsal fin, that Hardy Jones had named Chopper in his film. I met Blaze, a young adult female famous for the large white rectangular mark on her head, most likely a healed scar. Nippy, a mature female, became familiar to me, recognizable by the cut-off tip of her left fluke. Her female offspring, Pictures, identified by rake marks and other subtle scars, trailed along with her mother. A lot had happened in six weeks.

Since I used mnemonics as a tool to label animals with their recognizable features, by the end of the summer I had many Gashs (Little Gash, Big Gash), spots (Rosemole, White Patches), and behavioral names (Romeo, because of his friendly nature). I not only took photos but I sketched the dolphin's unique marks in my field book. My sketchbook became a valuable tool in the field since there were no digital cameras at the time (it took weeks to get slides processed and back to review). Of course in the later years the digital age had appeared, making it easy to process pictures in the field. In the 1980s we took our film back to Florida in between trips, developed the slides, and went through the meticulous process of photo-identification with a slide table. Now my graduate students do this work digitally onboard the boat in the evenings or during days with bad weather. This new process saves countless months of lab work, the expense of processing slide film, and helps with updated identifications in the field. But in 1985 pencil and paper would have to suffice, and they did.

After my first summer it was clear to me that this was an incredible opportunity to study a group of wild dolphins in clear water and on a regular basis, something not easily accomplished in the open ocean. But it would need a grounded framework in knowing who the dolphins were in their own society and in this environment. The experience of being with them in the wild is so powerful that it would be easy to make unfounded speculations and assumptions. "Twenty years," I thought, "that will be my minimum investment in fieldwork." I wanted to follow them through their long lives, follow their relationships, see who associated with whom, and understand how and why they communicated with each other.

The water is not our element, and glimpsing into their world is difficult. Keeping up with dolphins while they travel is not easy for a human swimmer. The dolphins can disappear when they want to. I knew of other friendly dolphins in the world; lone bottlenose dolphins in scattered locations. But here in the Bahamas was a family of dolphins that were interested in humans and available to observe underwater. Here was my life's work. I've always wondered why no serious scientist was studying these dolphins when I journeyed there in 1985. Later I realized that the main reason was because the dolphins were friendly, thereby risking contamination of objectivity and therefore of no possible interest to science. Would we really see their natural behavior on a regular enough basis to understand them? I thought so, but I would have to prove it. I believed in the concept of working with the dolphins, in their own world, and on their own terms, once again leaning on the strengths and proof of previous primate work and anthropological field studies. This was not a traditional way of studying dolphins but I felt it would be productive and eventually provide a new framework for working with dolphins in the wild. Not to mention that we could observe their behavior underwater. But it would require trust on both sides.

There was clearly dolphin etiquette to learn, simple things, but things that mattered. When I caught up with a traveling group the dolphins would open a space for me and allow me to move along with them. First, the dolphins "positioned" me accordingly in their group, but when I tried to "change" my position, as in the case of trying to get a photograph, it was met with jittery turns, glances, and all-out breaking up of the dolphin formation. As I swam along with the group I watched the unfolding etiquette and patterns. Mother-calves and juvenile-juvenile would pec-to-pec rub excitedly as they met up, or observed something novel in the water. Rapid patty-cakes (back and forth rapidly repeated pec rubs) would gradually subside as mother and calf swam off. Farther ahead a trio of dolphins engaged in a pec-to-head rubbing event. One adult, slightly below and behind the other two dolphins, was receiving a pec rub on his head from each dolphin to his side. Trio pec rubs often occurred after fighting as potential peacemaking gestures, possibly analogous to chimpanzees reconciling by grooming. This clear and purposeful signaling system involved pectoral flipper rubbing related to the relationship of the dolphins involved as well as the body part rubbed. This was going to be a complicated communication system to decipher!

My first summer in the Bahamas had a very powerful impact on me—a feeling of being a student in the dolphin's classroom, a classroom of experience and experimentation. It dawned on me that I couldn't build a "blind" underwater to observe these dolphins; we could not just be observers, but instead there would be interaction and mutual curiosity. The key to this study would be to "be" in a relationship with the dolphins: with respect, clarity, empathy, and a critical observational eye for whatever they chose to show us. If they were willing to invest in a relationship, or at least tolerate our presence, then I was willing to invest a few decades into nonintrusive observations. My hope was to blend in harmlessly and observe their lives with each other, not as an intruder but as a friend somewhat familiar with their culture.

Dolphins have evolved very parallel, yet different, cultures to be explored. Just beneath their raw behavior and individual skin patterns are individual minds and personalities unique both to their society and likely, to this solar system. It is that depth that fascinated me when I was twelve years old and continued to do so this summer. Layer the dolphins' daily patterns, routines, feeding, courting, and fighting behavior with a complex group of individuals, and you get a species living in a choice filled, fun-filled, challenge-filled, fully actuated life. This is the life of a wild dolphin. It is a life we knew little about, being aliens ourselves to such an aquatic environment, but one I wanted to explore.

In the fall I returned to San Francisco. I was still in graduate school in California but it was clear where I would be spending my future summers. My path for the next few decades couldn't be clearer; I had found my life's work, and it was in the Bahamas frolicking in the waves and foraging on the sandy bottom.

1986—Marks, Sharks, and Barks

I spent the winter in San Francisco finishing my graduate work. The contrast between studying captive dolphins at MarineWorld and my summer in the Bahamas was extreme. Although I was learning observation skills and began to understand some basic dolphin behavior in graduate school, I was even more committed to seeing the real deal in the wild and observing a healthy wild society. Arriving back in Florida the following summer was a thrill and a shock at the same time. It was great to be working with familiar faces from last field season. Dan Sammis and Ro Lotufo, and new crew members Jack Kelly and Dave Schrenk were on board. But the *Dolphin* was a mess, having been untended all winter. Decks needed painting, sails needed tending, and the inside quarters needed some cosmetic work. Captain Dan had two weeks to fix her up and get her ready for a summer of fieldwork. I was to stay onboard all summer, not just six weeks. We would go back and forth, between weekly trips, all summer.

Getting used to the summer climate in Florida with its breathtaking humidity is hard, so we were all glad to be on the water again. Late May brought decent weather as we sailed over the deep blue liquid of the Gulf Stream. West End greeted us like the previous year, a sleepy little village marked primarily by the Jack Tar Marina, an old run-down but lively pseudo Club Med that hosted drunken Texans and Canadians. May 27 was our first day of the season looking for dolphins and we placed bets on which dolphins we would see first. I thought Nippy and Pictures and Dan and Ro guessed Blaze. They won. It was a thrill to get in the water after not being there for six months. Did the dolphins remember us? Did they realize that we were here again or were we just opportunistic scenery? Their behavior on this first trip soon unfolded and I got the feeling that they were fully acknowledging our presence. During the next few days many of our new acquaintances from last year showed up, including Stubby with his cut-off dorsal fin, Rosemole with her telltale black spot on the top of her rostrum, Pictures and her mom Nippy, and old Ridgeway himself. Although they were still a bit hesitant to stay around us for long periods of time, it was a good representation of the dolphins I now had in my identification catalog.

On the way back to Florida after the first trip we had an encounter with some offshore pantropical spotted dolphins at sunset, yet another species gracing the waters off of Florida. As we slipped into the deep blue of the Gulf Stream with a large school of dolphins obviously not used to having people around, the contrast couldn't be clearer. The resident, friendly group of dolphins we had worked with for the last week was a sharp difference from dolphins in the Gulf Stream. Here it was really like meeting aliens. I, too, looked at them and wondered who was related to whom and how they spent their time. In the Bahamas, where we often talked through our snorkels and mimicked the behaviors, the dolphins showed interest. Here in the Gulf Stream the dolphins kept their distance and eyed us cautiously. It reaffirmed the unique and important opportunity that awaited me in the Bahamas, that of a species mutually curious and occasionally accessible to human researchers.

By June the dolphins were getting comfortable with us in the water. We saw them regularly as they came over to our anchored boat and stayed around long enough for us to get some photographs for identification. Sometimes we saw the dolphins at a distance and took our underwater scooter out to see what was happening. A desolate area most of the time, only a treasure hunting vessel, the *Beacon,* made home on the sandbank where we anchored. As June unfolded the dolphins started showing up with new calves. During one very early morning encounter, I got in the water only to be "checked out" by four older adult males while the mothers and calves stayed on the periphery. I got out of the water and while I was describing the event to a passenger I looked off the stern to see tail slapping. I got back in the water and saw that the same four old dolphins had returned with more mother-calf groups and now they let them swim close to us. Did they use the tail slap to get our attention? I had seen them tail slap to get our attention for a bow ride, for the attention of another dolphin, and for the attention of a rambunctious calf. Well, it had worked and they had our attention.

My main field season has always been from May through September during hurricane season, about one hundred days at sea divided into ten trips or so. In winter the northernmost sandbank we work on is exposed to the northeast winds that rack the Atlantic coast, making fieldwork difficult. It is also difficult to cross the Gulf Stream in the winter as north winds run opposite the Gulf Stream, which itself runs north, kicking up ferocious waves. In the summer predominant winds come from the southeast and with an island south of us our field site enjoys some protection. Historically, the spring weather is quite a bit rougher than the glassy waters that we get in July and August. The transition months of May and June, when predominant winds from the northeast switch to southeast, bring waterspouts and squalls our way. Births occur in early spring around March and April, so the spring is the best time for seeing neonates and mother-calf groups escorted by older males. July and August bring sweltering heat and calm seas, in between threats of tropical storms and hurricanes.

Extraordinary things happened on our trips, beyond the research, that gave me an intimate sense of the dolphins. Once our first mate, Jack, was towing a passenger back to the boat during an encounter because she was tired. As I videotaped Jumper, a familiar female dolphin, she suddenly stopped what she was doing and went immediately over to the swimmers, flanked them, and led them back to the boat. Another bizarre phenomenon occurred with Jumper later that summer. Tom, one of our passengers, had constructed a crown made out of sargassum, a brown algae commonly found floating on the surface, and was wearing it on the boat for days. Suddenly Jumper surfaced with some sargassum right on her head, clearly imitating what Tom had been doing all week on the boat. Jumper spy-hopped next to the boat, eyeing us on deck as she flaunted her sargassum crown. Had she been observing Tom all week at the surface? This would be one of many incidences of spontaneous mimicry that we would observe over the years.

Rosemole and Little Gash, two now regular female juveniles, often played with Ro and me in the water. Today they had acquired a little filefish and were playing a game of keep-away with the stunned fish. The dolphins carried it ever so gently in their mouths and dropped it, inviting us to grab the terrorized thing. But right before one of us reached the poor fish, the dolphins showed their aquatic superiority and swooped in to grab the fish. This went on for thirty minutes. We examined the filefish after Rosemole and Little Gash left and the fish, although terrorized, was alive. Over the years I saw many such instances of fish as toys, some lived and some died in the game. Some tried hiding in our swimming suits or between the video camera and our faces while the dolphins buzzed and poked around us. Uncontrolled giggles and laughter were heard on deck as the fish explored new hiding places among the humans. Half the time I felt sorry for the fish and the other half of the time I tried to get the fish freed up to give it back to the dolphins, as it seemed the courteous thing to do.

At the end of the summer I again started noticing big swollen bellies on the dolphins. Nippy's belly looked really swollen as did other mothers with calves, but since the mothers already had had their calves it occurred to me that it might be swelling from lactation not from pregnancy. Nippy was a different case because she was both lactating (with her two-year-old calf Pictures) and pregnant with a second calf. It turns out the spotted dolphins can be both lactating and pregnant at the same time. It's a heavy load for a female to nourish both a growing fetus and a dependent calf. Lactation manifests as a localized bulging in the genital area where the mammary glands are and pregnancy is visible as an overall growth in girth, detectable in the water after five or six months. Dolphin gestation is about one year, and by the end of this season we had a few females who were very girthy and likely to give birth in the fall. Other females' pregnancies were barely detectable, but the presence of a calf the next spring indicated that they had been pregnant in the fall. We now were able to detect and monitor the process of pregnancy within our field season (May to September) to predict next year's mothers. Although false pregnancies and lost neonates are possible, the majority of the time we verified a calf the following spring.

Sharks were also a regular and natural part of our field season. Although the shallow sandbank offers good protection and an ability to see a large predator on the bottom, the dolphins still have shark encounters and so do we. We had just finished a very interesting trip during which the same group of juveniles had come by for an early swim every morning. At the end of this trip we had another "grand finale," a term we jokingly gave to the fact that the dolphins seemed to know we were leaving and gave us a grand send-off. I have often wondered how they knew. Were they just good at getting our schedule down or tapping into our distinct cues of leaving? This day we had a two-hour encounter, long even by our standards. Rosemole was there with the other juveniles of her group. As I rested off to the side and on the bottom, enjoying the feeling of being in the water, suddenly, as I moved my legs to swim, I kicked something behind me. I turned around, thinking I had kicked a fellow human or dolphin and I saw an eight-foot bull shark behind me. As I rose to the surface, calm but shaking, I announced to my crew, "I just got bumped by a bull shark and I am going back to the boat." I proceeded back to the boat and that was the last time I wore a neon green dive skin during low light, a likely attractor in the darkness. After I got out of the water we continued our journey south to head home and there on the bow appeared Rosemole. She rode the bow, squeaking and eyeing me, until I couldn't see her anymore in the gathering twilight. Did she see the bull shark sneak up behind me? I certainly didn't expect the dolphins to protect us from sharks, although that is a common myth. They have their own concerns and safety to think about during these events. But I couldn't help wondering if she had watched the incident. Later in the summer Rosemole herself got attacked and bears the scars still. I remember her coming by with her fresh wound visible from the distance. She approached the boat alone with a piece of sargassum. She swam around me a few times with her now-familiar whistle as I got a good look at her wound and took a picture. Then she headed down to the bottom to rub her wound vigorously. The next few days Rosemole was very distant, something we noticed with other wounded dolphins. Her friend Mugsy also showed up with a wound, and they both came by the boat together regularly enough for us to see their healing process, which was incredibly fast. Over the summer Rosemole eventually healed and began happily associating with her normal friends, but it was painful to return to Florida after this trip, not knowing whether Rosemole would be okay. I had a feeling we were both a bit rattled by our shark encounters.

The dolphins act nervous when there is a shark in the area, especially if they've been bitten recently. Luna whistled madly when a nurse shark appeared on the bottom during an encounter. She rounded up her two-year-old calf after erratically swimming and whistling and they swam away. Normally dolphins don't react to a nurse shark, but Luna was hypervigilant after her own recent experience of a severe bite to her body, possibly sustained while protecting her young. That day her vigilance was up and she was taking no chances.

Before the end of our 1986 field season we had a few more surprises in store. Both speak of the sensitivities that dolphins have and the awareness and intent behind some of their actions. The first involved the death of a passenger in his sleep on board the boat, most likely from a heart attack. This trip was doomed from the start. First the boat was not ready to go and Dan thought we should cancel the trip. We decided to go later because of weather so I drove to the airport to pick up our passengers so they could cross the Gulf Stream with us instead of fly over. It was fortuitous as Trans Air, the scheduled airline, had gone bankrupt overnight and would not be flying anyone to the Bahamas that day. So we crossed to West End, passengers aboard, but the next day was the strangest day of the whole summer.

We had just snorkeled on a shipwreck and were heading up to the dolphin area. One of our passengers went below to take a nap in his bunk. As we approached the wreck dolphins greeted us but they acted very unusual, coming within fifty feet of the boat but not closer. Captain Dan kept inviting them to bow ride by starting up the motor but each time the dolphins kept their distance. I remember distinctly staring dazed from the bow, talking to Dan about how strange they were acting and "what should we try now?" Dan decided to get in the water and as he swam over to the group Chopper swam overhead and then left quickly, leaving Dan in his wake and even more confused. It was then that we discovered our passenger had expired in his bunk and began consoling his wife and daughter. Could the dolphins have sensed something strange on board? Many of us have had strange experiences with dolphins or other marine mammals. Alexandra Morton, in her book *Listening to Whales,* describes thinking of a behavior right before a killer whale (*Orcinus orca*) mimicked the exact behavior. Could the dolphins have a keen sense that we are unable to tap into ourselves? Whether it was coincidence or circumstance, we headed back toward port to deal with the new priority of sad family matters. As we turned to head back south, the dolphins came to the side of our boat, not riding the bow as usual but instead flanking us fifty feet away in an aquatic escort. They always rode the bow or just left, but now they paralleled us in an organized fashion. After matters were attended to on land, we once again headed up to the dolphin grounds to finish our trip. The dolphins greeted us normally, rode the bow, and frolicked like they normally did, and we finished the rest of the trip without incident. Twenty-five years later I have never again observed the dolphins escort our boat in the same manner.

One early morning, while I was drinking coffee on the bridge, Dan joined me and noticed we were adrift in the deep water instead of anchored from the previous night (clear proof that I do need caffeine to wake up!). We had drifted four miles into the deep water and large boat traffic, so we started up the engines and headed back to the shallow sandbank. Along the way we picked up various dolphins, including Blaze, on the bow. As we watched her ride I looked down and saw our anchor and broken anchor line go by on the bottom. As the boat continued gliding forward under momentum, Blaze left the bow and headed over to the anchor and circled it until we turned our vessel around, launched the Zodiac, and retrieved our lost anchor. It was a nice interspecies gesture.

The summer was filled with both dolphin and human events. In this remote area of the Bahamas we often answer distress calls from boats in trouble or relay to the Coast Guard about such activity. Historically this area was a drug-running area for many a pirate, including modern-day ones, and we had already seen our share of high-speed chases and drops from planes. The Fourth of July fireworks that occurred at sea, usually on our boat or another dive boat in the area, could be mistaken for distress flares if it wasn't for the holiday. During slow days we explored the reefs and shipwrecks in the area. Traveling by night, we sailed under the stars, tracking our celestial dolphin, the constellation Delphinus, in the sky. On rough weather days we ran for cover, sometimes outrunning a giant waterspout, essentially a tornado on the water. There was rarely a lack of excitement between the weather and the dolphins.

As the end of our 1986 season came to a close we decided to have a "crew" day before heading back to Florida. Stubby, Romeo, and Little Gash all came by for a last encounter. It was early September and my thirtieth birthday. For once I didn't care about my cameras. I was in the water just swimming along with the dolphins for a change. Just as we were talking about heading back to Florida, Rosemole showed up. Of course the summer wouldn't have ended properly without a final grand finale à la Rosemole. This was the first time we had seen Rosemole with her normal complement of companions since her shark bite incident. We kept getting in and out of the water and then finally we realized that the dolphins just wanted to bow ride. One by one the dolphins that rode the bow dropped off, until only Rosemole was left on the bow. Then with a turn to the side she was gone, and we seasonally migrant humans continued back to Florida for another winter on land.

As I reflected on my second summer with the dolphins, I realized how well the strategy of anchoring the boat in one place had worked out. My field site is quite large, about four hundred square miles, and encompasses a variety of depths from six feet to the deepwater edge and habitats including reefs, white sand, and grass beds. The two hundred or so spotted dolphin community is divided into three distinct groups that I called north, central, and south. In the early years I anchored the boat in a central location and just waited. It probably helped the dolphins feel comfortable, knowing when and where to find us on a consistent basis. It also gave them the choice, and us a clear signal about their level of interest. My main strategy of contacting the dolphins was to simply be available in their environment. This summer also had a different feel to it than last summer. Going through the entire season, the weather, the different dolphin behavior and patterns, was important. Our relationship with them had developed, strengthened, and changed over the summer. The other aspect that surprised me was my personal relationship with Rosemole. She was definitely the dolphin I had connected with the previous year. This time when leaving, it felt like leaving good friends. Last year it was like meeting and then leaving someone you really wanted to get to know better. This year I felt like we were actually part of the dolphins' regular routine, thoroughly incorporated into their system. There was also a real sense of the mutual process of discovery between two species—humans and dolphins. Being out on the boat, in the water and on the ocean, had sharpened my senses for observation in the wild. But I was still left wondering what the dolphins think when we leave and don't show up again for eight months. Perhaps to them it is us who are the seasonal migrants.

Redocumenting markings on individual dolphins, over consecutive years, has always been critical to my long-term work. With spotted dolphins, I tracked whatever marks were visible, including gashes, spots, and coloration patterns. Some marks were permanent, such as gashes or large shark wounds. Some markers were seasonal, yet were helpful in identifying dolphins within a week or season. These marks included rake marks from other dolphins, small puncture wounds, and small stalked barnacles that attached themselves to the dolphin's flukes and fins, at least for a time, and helped verify an individual. Even various sizes of cetacean-specific remoras were useful as seasonal markers. Although we see other remoras free-swimming, including "shark suckers," the species that attach themselves to sharks and turtles, I have never seen this species of remora free-swimming. The top of the head of a remora resembles the bottom of a tennis shoe complete with ridges and indentations (it is actually a modified dorsal fin). Remoras hold on to their host with suction while looking for scraps of food or productive water in which to feast. Over the years I had noticed that spotted dolphins with injuries or skin diseases often have these remoras attached. The remoras look extremely annoying to the dolphins. They meander their way around the dolphin's head, mouth, and belly, keeping just enough suction to not be thrown off. I have seen dolphins leap repeatedly in the air, slamming their bodies down until their bellies were bright pink from hitting the surface. Yet the remoras cling on tightly. I have seen a mother dolphin try to chase and nip at a remora on her calf, but the remoras, when threatened, cling even tighter to their host, leaving a telltale tennis shoe mark. But for all their potential annoyance, I have come to understand that the remoras are actually helping the dolphins.

In the lab we tracked melon marks, fine lines of color extending from the blowhole toward the rostrum and back to the eye, and throat marks, both quite distinctive to individuals. In one case, a previously lost individual was reidentified ten years later by matching a melon mark. There were long-distance markers and close-up marks. The overall body coloration of spotted dolphins can be quite distinct even at a distance. Some dolphins were labeled "Neapolitan" because the midline on their sides delineated a clear dark upper body in contrast to a white lower body. Other dolphins had spots that blended from top to bottom. Some dolphins, such as Little Gash, had very few spots on their ventral sides even into adulthood. (This might be one of the reasons why in studies of pantropical spotted dolphins, there is such variation in true "age class," verified by aging teeth versus verified by "coloration" categories or degree of spotting.) Some of these individuals may just have the genetics for certain degrees of body spotting or coloration. All of these markers helped me identify individuals at a distance.

There were close-up markers I used for times when dolphins were swimming nearby. Flying A, named for her "brand" in the shape of a sideways *A*, could not be mistaken. Even in 2007, I could still see it underneath her multiplying spots. Bumper, a northern female spotted, was named for her odd "bump" on the right side of her tail. The bump seen close up and at an angle clearly identified her.

I chose to name our individuals, rather than give them numbers. Not only does it make it easier in the water to, in your head, trigger a name, but like Jane Goodall and other long-term researchers, I believe nonhuman animals have personalities and should be given names. In the water it is necessary to quickly identify individuals by their marks in order to track behavioral activities. Later, as relationships between spotted dolphins became clearer, I began naming offspring with the first letter of their mothers, or sometimes in themes. Rosemole's first offspring I named Rosebud in 1991, followed by Rosepetal in 1994, and Rosita in 1999. The Snow family is led by Snowflake, a grandmother, and includes her offspring Snow, Sleet, Slush, Storm, and grandson Sunami (without the *t*). Paint, on the other hand, has a first offspring named Brush, my first attempt to try the "theme" idea. This prompted many interesting brainstorming sessions to find names; sometimes we resorted to the dictionary, cookbooks, whatever we could find to help us. Occasionally I would name a dolphin after a person, such as Caroh who is named for Carolyn Hay, an original board member of the project who succumbed to cancer in the late 1980s. My field notes became a bit complicated if I was describing dolphins and humans in the water, especially if they had the same name, so I tried not to use human names.

My "sketchbook" was now a valuable part of the fieldwork since slides were long in coming for identification at sea. I sketched out marks on Little Gash's "page." Some were new wounds, others gashes or rake marks, all used to track the subtleties easily detected by the eye. Even melon marks, distinctive and critical in the identification of young calves when they have no spots, were sketched. There really are some things that our eyes see better than a camera or the underwater video, and I often relied on these sketches over a season to help me identify individuals underwater or from the bow looking down. I remember even Dr. Randy Wells was amazed at this when he joined us in the field on one trip in the 1990s. After I had identified Ridgeway and Big Gash, two adult male dolphins on the bow, Randy expressed his amazement at identifying a dolphin so far from land. But for us, we knew what part of the sandbank we were on and we knew these dolphins and the constellation of spots and wounds on the tops of their heads. Even though most of my students don't use or rely on a sketchbook, I've noticed, now that I delegate the writing of encounter notes to some of my students, that it is quite natural, and apparently helpful to them, to sketch little dorsal fins and marks when coming out of the water. It is perhaps a natural process denoting such subtleties, and one that perhaps they have also seen the value of over the years.

Like 1985, this second field season also had a very powerful impact on me. I was seeing clear patterns to the dolphins' behavior and I learned the importance of using proper dolphin etiquette in the water. After all, we were in their world, and we had better learn their rules if we wanted to blend in. You can only learn dolphin etiquette from a dolphin. Is it impolite to approach a dolphin head-on? What does this signal mean in their own communication system? These experiences only reinforced my commitment to be out here for twenty years, a minimum time to learn the social signals of a multigenerational culture. I tried very hard to never push the dolphins beyond their comfort zone with our presence in the water, even at the expense of not getting a little more information. If it meant losing an identifying shot, then so be it. I would get one another time, but if I frightened them in any way, they might not come back at all. It was more important for me to have their trust than an immediate photograph of their marks, although both were welcome. It was an investment in a relationship and trust for the future that would, over the years, pay off handsomely.

Two types of dolphin encounters were also clear. The first was interactive: the dolphins made eye contact, broke from their own behavior, and came by to buzz us or interact with us, soliciting games or further eye contact. The second was observational: the dolphins engaged in their own behavior and we blended into the ocean as benign observers and documented their actions but kept our distance. During interactive encounters I used the opportunity to establish rapport and trust. This process was necessary to get the dolphins comfortable and familiar with us in the water. In the process I often got identification marks or sexed them. Soon my catalog of individual dolphin profiles grew to more than a hundred names, although there were still some dolphins to be identified. During observational encounters I videotaped their behavior, noting individuals and patterns of interaction that allowed for later detailed analysis of body postures and vocalizations. This process and information was the basis of the next twenty-five years of work in the field. And to this day, etiquette remains primary in our interactions with dolphins in the water.

As I grew to learn dolphin communication signals, the subtleties of such interaction became more and more challenging. Our human responsibilities in the water also became clearer. The dolphins had their own sense of which human beings were responsible for our activity in the water. For example, young calves are often very excitable and exploratory. Calves often seek out humans even when their mother is trying to get them to move or go somewhere else. In this context, the mother, if she knows us well, directs her communication signal, in this case a tail slap, toward a human instead of her calf, theoretically to signal "it's time to end this interaction—we dolphins need to move on." This is a very sophisticated sense of "other"; that the dolphins would choose to direct a dolphin signal at a human and expect the appropriate response. It is unlikely that they would mistake an awkward human in the water for a dolphin or dolphinlike cousin. It is possible that the dolphins are truly viewing us as a nondolphin form of a conspecific (another dolphin) with all the responsibilities and rules that come with it.

After my first full summer in the Bahamas I returned home to San Francisco and focused on my graduate work. I spent hours in my basement apartment playing back sequences of the captive dolphin vocalizations I was studying for my graduate work while my cat Kashmir, vying for attention, willingly mimicked the sequences of squawks with enthusiasm. Would I find the same types of sounds in the wild spotted dolphin behavior? Would they mean the same thing? I worked, I studied, and I wondered through the winter how different wild behavior would be from behaviors I saw in a tank. And I dreamed, sometimes while in a waking state, of the clear blue water and swirling dolphins.

~ 2 ~

Getting the Drift: Dolphin Behavior and Communication

> Any observer is an intruder in the domain of a wild animal and must remember that the rights of that animal supersede human interests. An observer must also keep in mind that an animal's memories of one day's contact might well be reflected in the following day's behavior.
>
> —Dian Fossey, Gorillas in the Mist

1987—Deciphering Dolphin Communication

In the summer of 1987 OSE had decided to charter a different boat for their ecotourism trips and I was not happy with the choice. *Bottomtime II* was a typical dive boat and held twelve passengers, good for OSE, but bad for research. And I keenly felt my challenges as a young female researcher on this trip. The captain, although outwardly friendly, was obstinately resistant to taking orders from a female. If I asked him to go southeast he would go northwest. It was not uncommon for me to have to call OSE from port to have a "chat" about the captain's behavior. We found ourselves in port quite a bit. It was also the captain's tendency to head toward Freeport, land of casinos and alcohol, with weather as an excuse. By the end of the field season I had adamantly communicated how much this vessel was not working for the research or the passengers. So it fell to me to find an alternative vessel for the next field season.

There were many pleasantries that summer despite the personal challenges. I met Chris Traughber, a long-term board member and friend over the years. Chris, much like my good friend Linda, puts his money where his mouth is. A longtime supporter of developing artists, Chris had always been interested in dolphins and communication. After boarding the boat he directed my attention to the newspaper ad for the OSE trip to be with wild dolphins, explaining that he didn't think it was possible and had to find out. Chris has been out almost every year since then, helping as a talented underwater photographer and vital member of the research project and our ship's medical doctor. I also met Gini Kopecky, Peyton Lee, and Lisa Fast, who became long-term supporters of my work through the years.

May 17 was our first day up on the sandbanks and Rosemole and Little Gash were the first to greet us. Rosemole's shark bite from last August had completely healed, blending smoothly into her dark skin, the scar barely visible. The pressures and complexities of having groups of people in the water were both subtle and surprising. Because we had twelve people this year (too many to be in the water at the same time), I decided to alternate swim teams during encounters. This often broke the continuity of the encounter and the dolphins left as a result. Yet we managed to have good and regular encounters with the dolphins. Rosemole and her juvenile friend Mugsy escorted us back to the boat sometimes as we drifted too far away from the safety of our anchored vessel. From the deck it was easy to see what was happening in the big picture and it was clear that sometimes the dolphins led people away from the boat, possibly to show them something of the dolphin's world, or as if a great treasure was to be found in the blue yonder.

But there were other reasons the dolphins brought us back to the vessel; one was a shark. In late July we had just finished observing a spotted dolphin breaching in a circle around the boat, following a frigate bird. I had just prepped the Zodiac to go out when the dolphins came to the boat following a horse-eyed jack school, the local fish species that chased small baitfish. Romeo and Ridgeway zoomed by, but suddenly stopped as if taking an interest in us. Ridgeway quickly left and we heard a series of modulated whistles in the distance. In the past when I had heard these whistles the dolphins had left rapidly. Instead, Romeo sank vertically down and calmly lay on the bottom. Richard, a passenger and adept free diver, joined him on the bottom, mimicking his behavior. From the surface I watched as a six-foot bull shark approached from the distance and headed directly toward Richard's fins. Romeo didn't move. Just as the shark was at his fins, Richard kicked and headed to the surface while the shark spooked and swam the other way. Then Romeo swam off slowly. Were the whistles a warning to Romeo? To us? Was Romeo's presence a signal to the shark, or a safety for us? Was he protecting us by remaining motionless? We would never know for sure. Perhaps he had never seen a human react to a shark and was looking for a little entertainment.

People often ask me how dolphins react to sharks. It really depends on the type of shark and the individual dolphin's situation. Mother-calf nursery groups tighten up, drop to the bottom, and swim away. Juvenile dolphins buzz and harass hammerhead sharks. Older dolphins monitor sharks, keeping distant but with an eye to the shark. I am convinced that dolphins are keenly aware of when a shark is on the hunt, and when its belly is full. It is not unusual to see a large tiger shark swimming among the spotted dolphins, with no apparent concern on the dolphins' part. Other times, the dolphins flee after a single scout returns from the direction of an approaching shark. Bull sharks are the most dangerous, followed by tiger sharks. Hammerheads, who eat stingrays, seem only to be target practice for the young dolphins. In the spring, large tiger sharks are seen following groups of dolphins with calves. Like everywhere in nature, predators go after the young, the inexperienced, and the weak. And it is no different in the Bahamas.

That's why discipline is very important in the dolphin society. Setting boundaries and rules for young individuals is a matter of life and death.

Calves are born with a variety of personalities (some are shy and some are bold) and can sometimes cause their mothers grief. A rambunctious calf is held down on the bottom by its mother as punishment for some misbehavior or transgression until it stops squeaking. Male calves sometimes seek the shelter of old male coalitions while the mother swims inverted in a rapid chase after the youngster.

This year I saw the first indications of teaching by a mother to her calf. Gemini, a now familiar fused female, and Gemer, her female calf, were scanning the bottom and feeding. Gemini went down, echolocated in the sand, and then glanced at Gemer who was also echolocating in the sand. They both poked and prodded the bottom with their beaks trying to scare up fish beneath the sand. Dolphins have good hearing and are good listeners. They also have the potential to "hear" the reflected echoes of dolphins scanning in front and to the side of them, eloquently measured by Mark Xitco and colleagues in captive dolphins. Later, I watched as a young calf went over to Stubby and observed him digging in the sand. The calf, Baby, went over to her mother, Blotches, and positioned herself above and a bit to the side and observed mom hunting, a perfect listening position to hear her mother's reflected echolocation clicks. Years later one of my graduate students, Courtney Bender, published results showing that mothers spent more time chasing a fish when their calf was present versus when it was not present. These data clearly showed that the mother is teaching the calf how to fish and is a good example of testing a qualitative observation quantitatively.

Maturing young females tend to younger calves in the group and take charge of them while the mother is nearby feeding. Babysitters are usually female but sometimes male. One midsummer day as we idled along, White Patches, Mugsy, and a third adult rode the bow. Suddenly they tail slapped twice, chuffed (a loud expulsion of air at the surface), and then drifted off. Thinking that this might be a signal to ride the bow we started up the engines. Mugsy then did an inverted tail slap, positioning her body upside down at the surface and slapping her fluke on the top of the water, creating another loud clap of sound. Did the intensification of such a signal convey a different meaning? It was as if we didn't get the message the first time and it was necessary for Mugsy to "yell" louder. Suddenly everything made sense. This observation sat well with what I had been observing with the babysitters for the last few years. When young females babysit, they tail-slap at the surface, and this signal brings her young charges over for protection or departure. The intensification of a tail slap is like yelling louder to get your point across.

Learning about discipline is obviously critical for young calves for future babysitting roles as well as immediate survival. Now that Little Gash was nearing sexual maturity she was babysitting young calves frequently. As I watched Latitude, Luna's two-year-old calf, erratically swimming out of control and excitedly vocalizing, sounding much like a squeaky rubber tire, Little Gash tried unsuccessfully to calm him down. In the end Luna came dashing in from the distance to take control. Apparently every mother has her limits and it is no different in the dolphin world and the skills of babysitting are likely learned through a combination of observation and participation. Males also participate as babysitters and disciplinarians, but not as frequently as females and not as skillfully.

Dolphins are also very visual animals, and for those that live in clear tropical waters their vision is excellent. I would venture to say that in the Bahamas dolphins use vision and passive hearing as primary senses at least during the day (in contrast to active acoustics). Dolphins, because of their streamlined adaptation to water, don't have facial features so many of their cues may not be initially obvious to us. But they display typical mammalian postures including head-to-head stances during fighting, body charges, and open-mouth displays.

Touch is clearly another important sense to dolphins during both social behavior and resting. As I watched Luna take charge of her calf, a gentle touch to the head had a calming effect. Nearby, Little Gash and two teenage dolphins indicated their excitement and surprise with a mutual and rapid pectoral flipper rub after a fish popped out of the rock on the bottom. Later that day I watched some adult males drift lazily along in the water column with their eyes closed. Because dolphins are voluntary breathers they don't sleep, they rest. And these four males were clearly resting. They swam along with their eyes closed and blinked only occasionally, as if the sun were too bright. As I clicked my Nikonos camera, one of the males, Mask, opened his eyes wide in surprise. Researchers have found that dolphins switch brain activity from one hemisphere to the other, alternating one for rest and one for activity, without rapid eye movement (REM) but going into theta waves, the meditative state of our brain. The opposite side of their body becomes hypersensitive to small changes in water pressure and hearing, likely adaptations to sense an approaching predator or trouble. Apparently the sharp click from my camera was enough to warrant investigation.

Sexual stimulation is common in many mammals and like some mammals dolphins masturbate. Males masturbate lying on the bottom with their penis in and out of the fine sand. Females masturbate by rubbing their genital area on the sandy bottom. Sometimes they get help from friends with pectoral rubs or beak-genital stimulation. They likely masturbate for one simple reason: it just feels good. Dolphins are like pygmy chimpanzees: they enjoy sexual stimulation across sexes and ages. Male coalition members are sometimes observed copulating with each other, perhaps as a sign of dominance or perhaps as a rite of passage into a coalition. Even male calves are encouraged by their mothers within the first month of life to copulate with her. Old males can aggressively chase and discipline calves, holding them on the bottom, during courtship chases of the calf's mother. For dolphins, sexual stimulation seems to be a natural part of the social code and bonding process. It truly is a sexual and social society.

As the summer drifted by I began noticing the dolphins "tasting" the water. I knew, from other research studies, that dolphins had a sense of taste, a chemical sense, but they didn't have smell. Dolphins have been tested for taste and can discriminate bitter, salty, and a few other things. But what would they need to "taste" in the wild? It was mid-July and the dolphins came directly to the boat in a very determined manner. There was a strong current and the dolphins hung right under our dive platform at the stern as they waited for people to get in, which they often did at this location where we jump off the boat. As I watched, Stubby approached Little Gash slowly, positioned himself near her genital area, and opened his mouth. Could the males be detecting chemical cues in the water for mating purposes? Could they tell when a female was in estrus? Next to Stubby, White Patches also inspected Little Gash with an open mouth in the water nearby. Of course open mouths can be an aggressive signal, but there were no other cues such as vocalizations or arched postures to indicate aggression. I was glad to know the individuals because over and over the scenario played out with different individuals, some male and some female. Little Gash visibly expelled urine and feces in the water and another male, White Spot, hung in the water vertically with his mouth slightly open, moving his head side to side. It's easy to understand why a male would want to assess the reproductive state of a female. Dolphins don't have visual cues, such as the swollen areas on female chimpanzees when in estrus, so dolphins likely use chemical cues. But why do the females inspect each other? Is it competition? Are they stimulating each other? It would not surprise me in the least. I have also observed these open-mouth, tongue-out behaviors when there were no other dolphins around, and the signals were not displayed toward us. Perhaps the taste lingers, perhaps they are following trails of some chemical cues, but I think they are "tasting" the water.

Late May rolled around with tropical storms already brewing in the Atlantic, preceding even the June 1 start of hurricane season. May 28 saw a tropical storm off Bermuda. On May 29, there was a new tropical depression east of Miami and we stayed in port until May 31 as the disturbance passed. Like the spotted dolphin age classes, tropical disturbances have developmental phases. A tropical wave (TW) is a clump of disturbance to keep an eye on. A tropical depression (TD), which gets a number like 1 or 2, is a more organized disturbance with sustained winds up to thirty-nine miles per hour. At forty miles per hour a disturbance is a tropical storm (TS) and gets a name and can grow in strength and organization until it becomes a hurricane at seventy-five miles per hour and greater. Hurricanes have their own categories from 1 to 5, one being the weakest and five and beyond being a nightmare. When we are working far from land any tropical disturbance creates a challenge for working in and on the water.

Early June came and went with more transitional weather and so we found ourselves in Freeport once again. We went to the Lucayan caves and saw bats, mangroves, and visited the new dolphin center at UNEXSO, the local dive shop. Mike Schultz was creating a new program and his dolphin arrivals were from the Gulf of Mexico. Watching these bottlenose dolphins taken from the Gulf of Mexico was a stark contrast to seeing a dolphin family frolic in the wild. After some convincing, and to their credit, UNEXSO's next batch of dolphins came from a facility in Nassau, Bahamas, that was shutting down; their tanks had substandard conditions and they wanted to get rid of their dolphins, so instead of capturing wild ones, UNEXSO rescued these individuals. I thought that was a win-win. Why not improve the home of those dolphins that can't be released and let them live their lives out in semidignity, at least being active in a program? Eventually UNEXSO moved their dolphins into a large lagoon with seawater exchange and trained them for open ocean dives with scuba divers, a better alternative than a sterile tank.

In these early years encounters with bottlenose dolphins were rare. Ironically, in parts of the world where dolphins are interested in people they are usually bottlenose dolphins. In the Bahamas the bottlenose dolphins were shy and cautious. I suspect this was the effect of repeated captures in this area in the 1970s for marine parks. One of the most amazing aspects of my fieldwork is the complex relationship between the spotted and bottlenose dolphins. These two species are sympatric, living and sharing the same environment. The first time I observed the two species interacting was in murky water, and for only a brief moment. I watched as the bottlenose dolphin stationed himself head-to-head with a group of spotted dolphins. The noise was overwhelming. "Too bad it was murky," I thought. "I might never see this again." But today I was watching Jumper (later renamed Dos), a well-known young adult spotted female with a bottlenose dolphin. Jumper was typically a mellow but curious dolphin. At first the bottlenose dolphin stayed on the periphery, out of sight, but then began swimming with one of the crew. People on deck said they saw six other bottlenose dolphins swimming in the background. Later, four bottlenose dolphins showed up, swimming in a family formation, and seemed curious, even without Jumper around. Then again in late August, Jumper showed up with more bottlenose dolphins, but this time the bottlenose dolphins were very young. A few minutes later Hedley, another female adult spotted, appeared with an old bottlenose dolphin. As these two separate but interacting species rolled and frolicked, I couldn't help wondering, given the shy nature of the bottlenose dolphins, if the spotted dolphins now familiar with us were bringing them to the boat. It was an interspecies introduction on many levels.

In early September we were out in the Zodiac following a group of traveling spotted dolphins. Young Apollo rode our bow with his mother Luna nearby. As we slipped in over the side I saw Apollo and White Patches (a well-known babysitter and eternal aunt) foraging on the bottom. Three new spotted dolphins and two bottlenose dolphins suddenly appeared, chasing, tail-biting, and displaying head-to-head postures with squawks and open mouths to one another. One bottlenose dolphin with his penis out was attempting to mate with the juvenile male spotted dolphins. What was going on here? This was crazy. Bottlenose dolphins were biting bottlenose dolphins, spotted dolphins were biting spotted dolphins, and bottlenose dolphins were biting spotted dolphins and trying to copulate. Was this a dominance display or an individual argument? Did the bottlenose and spotted dolphins communicate with each other? Did they use the same postures and sounds? Did such signals mean the same things?

I had studied the "squawks" of dolphins during aggressive behavior in captivity. Squawks are discrete bursts of sound, essentially packets of clicks in rapid succession. Unlike whistles that travel miles, burst-pulsed sounds like squawks are for short distance communication. These sounds are typically heard during aggression and fighting and are correlated with overt body postures such as arching the body and head-on displays with open mouths. During more escalated behavior, squawks can become synchronized, much like the chant in a football huddle, providing the listener with a sense of the size and strength of the group.

Eloquent synchronization of their movements emerged as the spotted dolphins grouped up and gracefully coordinated their body postures and sounds to fit the bottlenose dolphins. High-pitched "screams" (overlapping whistles) and low frequency "barks" (similar to a sea lion bark) filled the water as the dolphins chased one another in aggressive postures. Screams and barks are heard only during extremely escalated fights, suggesting that most aggressive activity is moderated and resolved before it reaches such a level, but not today. Chirps and whimpers (soft, partial whistle sounds) dotted the edges of my hearing as the younger spotted dolphins tried to appease the aggressive and vocal adult bottlenose dolphins. The young dolphins had resorted to other options and submissive behavior to dissipate the escalating aggression.

Because of the complex nature of dolphin society these conflicts often break out either over mating opportunities or territory. Aggression can be individual against individual, group against group, or species against species, but it was usually about male competition. After the vocalizations decreased momentarily, I watched as the spotted dolphins faced off with the bottlenose dolphins in a head-to-head position, at first remaining stationary, then escalating into body passes and figure-eight movements, body hits, and open-mouth displays. As the whites of their eyes became visible, the spotted males began to synchronize their actions. I found myself in the middle of a group of fighting males as they swirled around me and leaped out of the water to land nearby. They were equally engaged with one another but still cognizant of my presence. I never had the feeling I was in the way or that they might land on top of me. All I could think of was how much sound analysis I had to do. It was an incredible and complicated example of interspecies fighting.

Orcas in the Pacific Northwest have specific groups with specific dialects. Yet they share some calls. Would this be the scenario here? But these were two separate species. If such regular interactions occur it would make sense to have some cross-species language. Perhaps they used the same signals but with more emphasis or intensity when they were in mixed groups than with their own species. Over the years I explored the level to which bottlenose dolphins and spotted dolphins shared signals including body postures and vocalization types. Although both species share some sound types, both bottlenose and spotted dolphins retain some of their species identity with unique vocalizations. There is nothing more fascinating than the relationship between these two species and the degree of intimacy they share in the wild, including conflict and cooperation.

The rest of the summer I continued my normal work. Photo-identifying the dolphins, developing rapport, videotaping behavior, and collecting sounds. Blaze and Paint, two females, were now familiar dolphins and clearly important members of the northern community of dolphins. The bottlenose dolphins in the area had become a bit more comfortable with our presence but I still had not seen any examples of their own social behavior such as play, fighting, or general socializing. As the summer flew by we went in and out of Fort Lauderdale for our trips, dodging waterspouts in the Gulf Stream as we ran from storms and tropical depressions. It was a turbulent fall, with depression after depression. By September 22 Hurricane Emily was heading our way from the east so we headed back to Florida. We ended our trip, and the field season, two days early. I was getting my first taste of a real hurricane season, and I didn't like it.

1988—Swim-Aways

During the winter I decided to venture back over to the islands to see if it was possible to work during other months with the dolphins. As luck would have it a cold front had descended on Florida. With an air temperature of thirty-nine degrees, and seawater drenching us, I decided that it might be more of a challenge than I had thought. It's hard to get sympathy for working in the Bahamas. Most of my colleagues have only seen video examples of behavior from the clear water and calm days. But fieldwork is fieldwork and if you think it's easy to work in cold air, getting in and out of the boat from the water, think again. So ironically, hurricane season (June through November) remained our regular field season, and even that was challenging.

This winter I was charged with finding a decent vessel for OSE to use during the upcoming field season and so I set my sights on a multihull for stability, either a catamaran or trimaran. Because we work far offshore and anchor much of the time, the stability of a multihull vessel is critical to comfort, human and equipment safety, and data collection. Although I found a trimaran that would suffice, the owner was unable to supply the necessary liability insurance required by OSE, and they chose another vessel for the summer. The *Jennifer Marie* was a steel-hulled, handbuilt monohull sailboat from New York. A tough ship in some ways, but ill-suited for open ocean research at anchor on the sandbank during hurricane season.

The spring of 1988 brought clear weather as we sailed across the cobalt blue Gulf Stream on our first May trip. There were quite a few boats in the area, possibly too much human activity for the dolphins as they were harder and harder to find. Late May brought ten-foot seas to our field site and a tropical depression off of Cuba, leaving us hiding behind the island much of the time. June, often a turbulent month, brought storm after storm through our study area. It was not uncommon for the dolphins to display elaborate tail slapping preceding a storm. There are anecdotes about sailors being directed or led by dolphins to land during storms or navigated through harrowing channels. I've often observed that the dolphins come up to our anchored vessel and tail-slap before a squall or strong storm moves through. I had wondered if the frequency or intensity of tail slaps has any relationship to the severity of the approaching storm, or if it is a warning to us, for them, or both. By this time I had seen many examples of how they use tail slaps, to get attention, to round up the calves, and to signal the desire for a bow ride. This June morning we awoke to heavy seas, a Beaufort 5. We watched, unable to get in the rough water, as the dolphins surfed off our bow and next to our heaving boat for two hours, tail slapping and frolicking in the large waves.

Most of my work had been with the central group, and included core dolphins such as Luna, Stubby, Little Gash, and Rosemole and her family (the Rose family). Next door resided the northern group including the Snow family, Paint and Brush, which I called the Underbite family because of their distinct lower jaws. Farther south, and a bit more elusive, were the southern dolphins, which include Venus, Flying A, and others. All three groups made up a coherent community of dolphins on the Little Bahama Bank and although they had specific home ranges they sometimes overlapped in their activities. The males had more extensive ranges, probably owing to the need to spread their genes among the community.

Human-dolphin interactions were continuing on a regular basis as we two species got to know each other and the dolphins usually did a better job at it than we did. One day we saw of group of dolphins breaking the surface of the water, exiting rapidly as they flew in the air toward our boat. I got in with the video with a few regular crew and some of our handful of ecotourism passengers (usually a total complement of ten to eleven people on the boat). There were fifteen to twenty dolphins surrounding us. I was holding on to nine-year-old Cathy with one hand and my video with the other trying to record Gemini, Gemer, and White Patches swimming around us. White Patches made some small whistles, infantlike chirps, as she circled around us. White Patches, the eternal babysitter herself, had never seen me babysitting a young human before. Her excitement vocalizations were audible and electric and she continued to swim around us, eyeing the human youngster attached to me. Later in that same encounter I watched Rosemole with a group of young males. It was a quiet affair with the six other juveniles grouped around us while these immature males went belly to belly with Rosemole, taking turns practicing their mating behavior. I felt like I was observing some primitive ceremony, a mating ritual. As we humbled humans watched, the dolphins opened up the circle and allowed us into the center of the activity. It was an invitation to watch this most typical dolphin behavior.

Of course play is an important part of the bonding process between mothers and calves, and between juvenile dolphins, and it takes a variety of forms. Play occurs during practice courtship behavior, during mock fights, and during play with objects like sargassum or terrorized fish. In addition to their natural toys dolphins occasionally find plastic trash in the ocean, a very disturbing and too common sight these days. In mid-July, we stole from Diamond, less than a year old, a piece of plastic she had adopted as a toy, dragging it around on her pectoral flipper and fluke. As we left the scene, feeling like thieves, we stopped to scoop up yet another balloon floating along, remnants from a cruise ship party where a family celebrated a child's birthday, unknowingly endangering Diamond in the process.

By this time it was clear that Rosemole, Little Gash, and Mugsy were fast friends. White Patches, the eternal aunt, was regularly mating with the subadult males, likely just for practice, but for the most part, White Patches was the dolphin equivalent of a tomboy. Although she was already a mottled female, and likely sexually mature, we had yet to see her pregnant or with her own calf. Horseshoe and his young juvenile friends formed a large group, probably a future coalition, and awkwardly chased and herded White Patches while the males jockeyed for position and clumsily approached her. Much like in human society, nothing in dolphin society is easily learned and this clumsy behavior was the mark of inexperienced young males. On the periphery Rosemole and Mugsy watched nearby as other males took turns and positioned themselves underneath Little Gash for an opportunity to mate. Sometimes Little Gash slowed her pace to cooperate, other times she dashed away or tail slapped her would-be suitors to communicate that she was not interested.

At some point play mating becomes real mating and serious dolphin courtship is very ritualized with a variety of outcomes. The males typically approach the female in an *S* posture (an extreme contortion of the body creating the shape of an *S*) while orienting their rostrums to her genital area. Then loud buzzing, a specific train of clicks, with very rapid repetition rates is heard. This sound is used in three contexts: courtship and mating (with the males buzzing the genital area of the female for stimulation), discipline (during which the mother holds a calf down on the bottom), and chasing sharks (when dolphins group together and orient on the shark's body using this sound). You can often see the shark twitching, so it is likely that the buzz has a tactile component. It is an intense sound, and such intense sounds are likely felt as well as heard.

Beak-genital behavior is a common behavior during which one dolphin puts its beak on the genital area of another. Scientists have speculated that this is a mechanism for the male to stimulate the female before mating. It can mistakenly look like nursing if a young dolphin and adult are involved. Most frequently male dolphins use beak-genital behavior during the stimulation and inspection of a female dolphin during courtship. However female-female beak-genital inspection is not unusual. This summer as I watched White Patches inspecting Little Gash this way, I couldn't help wondering whether the females can detect either health or sexual receptivity during this behavior. I had previously noticed various dolphins "tasting" the water nearby each other. I was once again reminded how important it was to know the sex of individual dolphins to make sense of an observation. This could have easily been activity between a male and female, but I knew them both, and I knew that in this case I was watching two females.

Dolphins love to have sex and they have sex a lot. Dolphins are like pygmy chimpanzees (bonobos), highly sexual across all ages before they are sexually mature, and they have sex for both pleasure and procreation. Young males and female dolphins practice in small groups, in large groups, basically whenever they can. Over the years I had wondered when, during the dolphins' life cycle, courtship becomes serious and actually leads to procreation. This summer I watched Romeo, one of the oldest males in our group, regularly associating with a trio of other old male adults. Even though this quartet was not involved in the ruckus and rambunctious activities that younger males often were, I wondered if Romeo and his male friends were more representative of the true mating opportunities of this group. In their approach to females, older males are more intense about their buzzing and mating, cutting to the chase. I had observed the extensive "ventral keel" of old males, likely from the distension of their testicles that double in size during male rut. Does a ventral keel give a visual indication to the female that a male is ready for mating? Males in rut look like pregnant females because of their extended ventral areas, but after time are distinguishable due to the location of the keels. I wondered whether the young male-to-male activity was practice for the formation of strong male coalition bonds more than reproduction. Sometimes these coalitions seemed to be about mating, but sometimes they were about male relationships. Ridgeway, a male coalition partner, was sometimes observed with unknown males in the area, and I wondered if these males switch, or continue switching, coalitions even when they're very old. But how do young males form coalitions? Do they join existing coalitions? Do they form new coalitions with their friends and then vie for coalition rank or opportunities? This summer I noted young Poindexter, and his young male friends Cat Scratches and Slice, had new rake marks, tooth marks made by other dolphins during play or aggression. I wondered if these friendships formed during juvenile years. I was just beginning to see young males in our group grow up but it would be another decade before the process of coalition formation became clear.

This June even the Gulf Stream was too rough to cross in our monohull sailboat from Florida. Every day we got up, every day we tried to cross to the Bahamas, and in the end we canceled the trip. My soon-to-be good friend Pat Weyer was on that fateful trip but had managed to talk the captain into allowing her to remain on board for the next trip as well. Pat is a famous glass artist from Seattle and early in 1988 she had contacted me about helping with some artwork from my project. We became fast friends, sharing an interest not only in dolphins but also in ancient Greek archeology. Pat stayed on the next trip to join a group from New Orleans, including Keith Twitchell and Thea Morgan who would also become board members and key supporters of the project down the road.

The summer flew by with now familiar dolphins and behavior to observe. In mid-July Dan and Ro came out on another boat to get married. As they approached the sandbank and hovered nearby I jumped over the side and swam over to join them for the ceremony. With the wind blowing and the dolphins on the periphery, they said their vows, truly a wedding at sea.

In early August, I began my swim-away experiments. It had become obvious that the dolphins were interested in leading us away from the boat on occasion, but it is a potential safety hazard. My guess is that they wanted to show us part of their world, or other dolphins, in the distance. Gini Kopecky, a longtime member of the project, was once led away by some dolphins only to remember to turn back to the boat as per our safety instructions for beyond one hundred yards, but as she looked ahead she saw a wall of twenty dolphins awaiting her. So I decided to develop a way in which we could safely let ourselves be led out to sea to see what awaited us. The swim-away teams were organized as a swim team, a kayak team to follow the swimmers, and the boat team to follow the kayak with the large boat. But the dolphins were the ones uneasy about us leaving the perimeter of the safety line that trailed behind the boat for a hundred yards, and they rapidly brought us back to the mother ship, or to the kayak. This was often right before a squall, which was lucky for us, and acutely wise by the dolphins. In some instances the dolphins preceded bringing us back with a tail slap and then a synchronized movement of the group. After they left the boat they tail slapped repetitively until they had rounded the other dolphins up to leave.

It was not so easy to drift with the dolphins under these ordinary conditions. If they came to the mother ship to interact, and we tried to leave, it broke spontaneity. When we allowed the dolphins to lead the encounter they still swam us in circles or brought us back to the mother ship. Later we were successful drifting with them for hours with the mother ship, getting in and out of the water repeatedly, as we drifted over reefs and grass beds watching the behaviors they engaged in while not on the white sand. Although the kayaks seemed like a good and safe way to stay with the dolphins, we eventually gave up on the swim-away plan. We learned to use the kayak for sunset paddles with the dolphins and listened to their squeaks and whistles through the hull of the kayak, just two species hanging out together.

Despite the boat challenges, this summer two other things were happening. The bottlenose dolphins were starting to get more comfortable with our presence in the water, probably from watching us observe the spotted dolphins during their mixed-species encounters. Secondly, I was learning about the individuality of some of the dolphins like Paint, who had her distinct style of greeting me, and only me, in the water. She approached, dove vertically, and hung in the water column until I mimicked her vertically. Then she swam around excitedly as if she had accomplished a major coup teaching this human something, or perhaps it was in the simple awareness that she was recognized as a unique individual. She was aware that I was aware of her signals and I found a way to communicate it by mimicking her actions. This became my regular greeting with Paint, unique to her and then incorporated by her offspring years later.

By the end of the summer it was also clear that Rosemole was becoming a reliable babysitter for young juveniles. By now I knew Rosemole well, along with her best friends Little Gash and Mugsy. Sometimes we saw Rosemole rounding up her younger counterparts, tail-slapping the surface and producing a whistle. Sometimes the youngsters went willingly. Other times it took Rosemole's physical presence to herd them together. Rosemole became less active with us, was now a regular babysitter, and had such an intense blooming of spots I could barely recognize her. I came to learn that both babysitting activity and rapid spot accumulation were good predictors of a first pregnancy. And learning to babysit was an important juvenile activity, making the dolphin's first experience as a mother more successful, as would be the case with Rosemole.

By mid-August the hurricane season was in full swing, and we found ourselves in port much of the time dodging tropical depressions. The *Jennifer Marie* was no match for the ferocious swells in our study site. The ship repeatedly crossed from Fort Lauderdale over to the Bahamas on our weekly trips. One mid-August trip we arrived in West End, only to find a bruised purple sky growing on the horizon. Some of the passengers from Texas and I were well aware of what a tropical depression looked like, but the captain insisted we head up to the study site. Five hours later we arrived, only to notice that the impending storm was now a smoldering cauldron of anger. Unwilling to head back to land, the captain decided to wait until morning, but what a morning it was to be with eight-foot seas pounding us at anchor. We headed back to land with the storm in full fury and now were trapped belowdecks in our bunks. We watched as buckets of water poured through the nonsealing hatch above. Vomit and Ziploc bags full of saltines, which the captain's wife had thrown down from above, floated by in a slick mixture on our inside lake for the next fifteen hours. The pale pea green hue of seasickness is rarely seen on my face, but even the weathered sailor in me was verging on that color. Some passengers walked off the ship when we arrived, never to return. Once again I was reminded of the fine line between life and death on the sea, and it was nothing to be lax about in the field.

In the field we are both helped and biased by our available technology. I remember one particular experience in my early years that showed how much we rely on and how bound we can be to our dependence on technology. I've been poor most of my life, and in the early years of the study I barely had enough funding for cameras and film. One year, by the end of the field season, my underwater video had died, and eventually my one camera, my underwater Nikonos, also stopped working.

Unable to afford to replace my equipment for the last two trips, I decided to use the time as best I could. I spent my time in the water and sketched out marks, sexed individuals, and observed their underwater behavior without the distraction of a camera. I considered the time productive and was able to increase my identifications and basic understanding of their behavior. Interestingly, the guests on board, who were OSE participants, had another opinion. They could not understand how I was "doing" science without any tools. Now I thought, "How interesting, I've always considered my brain a tool and my abilities to observe." Apparently they equated science with technology, but the process of observations in science is primary, and equally as important as collecting data.

In many ways science is like an art. In my mind there are two kinds of scientists. There are the "bean counters" who happily follow some previous method, rely only on numbers and statistics, and who don't really explore and "mine" their data or thoughts. It is not atypical for this type of scientist to not question things like whether the method is appropriate to the question or the results. Did we measure the right thing? Is there another way of asking the question, some other framework to consider? The second type is the creative scientist who is probably a minority. I think real breakthroughs only come from these original thinkers. These are the artists. Perhaps the first marker is that they tackle a new problem, often one that has been ignored either because no one thought of a new way to approach it or that it was deemed unworthy of science because of the current paradigm. First is the idea, then the discovery and collection of data, then germination of the ideas and data, and then finally the product, albeit often a paper one. In many cases, their keen eyes catch a mistake or anomaly in these data that actually points to the solution or process they are trying to reveal. In this day and age, funding comes strongly from the applied science side: "designate a problem and find a solution." What science is best at is illuminating natural laws, processes, and patterns. In his book, *The Structure of Scientific Revolutions,* Thomas Kuhn describes the resistance of scientists to overwhelming data and its implications in emerging and controversial fields. This is particularly relevant today with strong evidence of nonhuman animal intelligence and emotions. In the end resistance is futile, but it may take decades to accept new paradigms and ways of thinking.

Many of us, including myself, are aghast at the lack of natural history data and observations in the sciences today. Students want quick data, don't even know how to observe, nor are they willing to spend the time in the woods looking at how things really work. Some rely on the Internet for data, and reference blog pages as if someone's opinion is a substitute for reading an original research paper and applying some critical thinking. Big problem. I have seen it in graduate students at my own university. At the risk of sounding like my grandmother who reminded us at many intervals growing up that when she was young she had to "pick potatoes, peel potatoes, and mash potatoes," the hardships of another generation may be an appropriate example. The process of spending time with the natural world is in jeopardy, causing us to lose sight of it while at the same time we are trying to measure it.

I remember sitting on a bus, at a conference, with my colleague Dr. Jim Darling. Jim has researched humpback whales in Hawaii for years, including the issue of why male humpback whales sing on the mating grounds. On the bus I asked Jim what he was up to and he told me about his work in Canada. The Canadian government had suddenly realized that they didn't even know the basics of what some of the wildlife did in the woods, like where a bear goes and how it spends its day. So they were putting people out in cabins to observe. Jim was working with an Indian tribe in the area. He was describing how he could go down to the coast where his native colleague would describe how a starfish would be on this side of the rock during a particular tidal cycle and a different side in a changing weather pattern. The native people that lived close to nature knew it like no one else. There is simply no substitute for spending time observing nature.

The rest of the summer we dodged heavy seas at our field site. A monohull vessel is a great disadvantage where I work because it is just not stable enough to anchor in the large ocean swells and changing tidal flows.

This summer my data collection was half of normal, due to the fact that we repeatedly had to leave the area. Time spent traveling to and fro was time wasted outside the dolphin area and my data, or lack thereof, proved that. We sat in port waiting for tropical depressions to pass by the Abacos to our north and Cuba to our south. I parasailed for the first time for a little excitement, but I really wanted to be on the sandbank working with the dolphins. I was determined to get back into a multihull vessel for the future. It was better for the work, better for the equipment, better for a long-term project. On September 6, as the end of our field season drew near, Hurricane Gabrielle moved toward Bermuda, sending large waves from a thousand miles away but luckily impacting only us, not the dolphins. We rolled and crashed in the eight-foot storm swells invading the sandbank, causing breakers on the shallowest part of the sand ridge. This would not be the last time hurricanes would affect my fieldwork.

Both 1985 and 1986 had been incredible weather years. But 1987 and 1988 gave me a taste of things to come with dynamic weather patterns. In 1988 I got a taste of what the summer was really like (in non–El Niño years) in the Bahamas. My graduate students often wonder, if their first trip features great weather and dolphins, why we talk about the hardships of being in the field. They, too, are seduced into the pleasures of the work only to find that it is like every other field site, with its good, bad, and ugly days. Tropical storms and depressions keep us in port or, in some cases, have us heading back to Florida to escape developing hurricanes.

By the end of the summer I decided to strike out on my own without Ocean Society Expeditions as an umbrella organization for my work. It was clear, from various issues, that research was not their priority, but it was mine. I decided to form my own nonprofit organization. I dragged along Linda, Chris, and other longtime friends to create the structure that would umbrella the research for decades. There are things I probably don't even remember about these early years. There were joyous moments when we saw first-time behaviors, and there were major frustrations having to deal with captains and changing research platforms. After four years in the Bahamas it was clear that the dolphins were showing me their natural behavior and opening up the secrets of their world. I wanted to do research, and I wanted to be out there as much as possible to observe this aquatic society on a regular basis underwater. As we headed back to port on my thirty-third birthday I pondered the fact that I would need to get my own boat. I also realized that such a momentous move was scary as hell.

~ 3 ~

Realities of Life at Sea

Now I see the secret of the making of the best persons. It is
to grow in the open air and to eat and sleep with the earth.
—Walt Whitman, Song of the Open Road

1989—The Wild Dolphin Project Is Born

Big changes happened in 1989. I had packed my books, my cats, my car, and moved to Florida in the spring after finishing my graduate work in San Francisco. The previous field season had been painfully short, so I decided to focus my energies on obtaining a catamaran for the fieldwork. By this time I had also decided to split with OSE.

The *Wren of Aln* was a large sailing catamaran. Built in England, and named for the river she was built on, she was sailed by her captain across the Atlantic and his hand-built vessel served as his family's home for many years. Doug Newbigin had agreed to a three-year lease for our fieldwork. A beautiful sailing catamaran on deck, the *Wren*'s inside features left much to be desired. Built to accommodate about five short people, the *Wren* also had a wild bathtub that served as a shower, and a forward berth with only a deck exit that made sheltering from a storm difficult. She worked well as a stable research vessel and at the end of the trips we were salty and bruised but happy. Manipulating sails while working with the dolphins was challenging and took new strategies and extra time. On windy days we sailed around the sandbank with the stiff breezes powering our way. Carrying only four hundred gallons of water for ten days, we bathed in the ocean; we baked and rebaked in the scorching sun, and we bruised ourselves running into head beams and low ceilings. Although it wasn't anywhere near as comfortable as Larry's vessel *Dolphin,* it was good to have a catamaran again. After two years of different boats and captains I was able to hire Dan, our original captain, for the duration of our lease. Together with a technical assistant, Dave, and various dedicated cooks, we once again headed out to the Bahamas in May.

It was not only our first season on the *Wren;* it was also my first season running the boat under my new nonprofit, the Wild Dolphin Project. We were poor but we were at sea. With willing participants who helped finance the trips, and volunteer cooks and technicians, we embarked on a new phase of the work—our own. It was a good decision. Although it was a steep learning curve with many financial challenges, it gave me the flexibility to hire my own captain, design my own field schedule, and decide my own direction. It was the solution, albeit a hard one. Some ecotourism platforms can be used for research, but it was critical to have more control over the number of passengers and where the money went.

Before our summer field season started I had been out to the Bahamas in March on an unusual trip. My friend Patxi Pastor had arranged a trip with his friend Bill O'Donnell to go out and see the dolphins. Little did I know that March is historically the windiest month in the Bahamas. Bill had chartered a large yacht so with a stretch of good weather at hand we anchored in the dolphin area for a week. Bill's three sons, Ryan, Billy, and Christian, all under the age of twelve, were a joy and as well behaved as you can get in the water. The dolphins were curious, circling around them as if they had never seen "small" humans. Over the years I have actually been at our study site in every month, and usually some of the dolphins are around. Because of the weather challenges in the winter, it is hard to spend the same amount of time in our study area as in the summer. But it appeared that the spotted dolphins were indeed resident year-round.

This March had one big surprise awaiting us. White Patches who, like Rosemole, had become a reliable babysitter for young dolphins, showed up at the boat babysitting a young bottlenose calf! I couldn't, at the time, imagine anything stranger. It was the first hint of the complexity and intimate behaviors between these two species and I was seeing it for the first time. Occasionally the bottlenose and spotted dolphins fed in the same area, but it was rarely a mutual activity. When I saw White Patches with the bottlenose dolphin calf, I knew that this was no ordinary relationship. Over the years it became clear that young female spotted dolphins, just coming into sexual maturity, often babysit young bottlenose calves for short periods of time. Interspecies babysitting was yet another facet of this fascinating mixed-species relationship.

Stubbet and Barb, two northern females, became regular visitors with the O'Donnells. Stubbet, known for her extreme underbite, and Barb, with a fishhook-shaped dorsal fin, were a pair of troublemakers. Tricksters and jokesters, they snuck up behind the kids in the water and gave them an occasional startle. Then both females surrounded the boys with the gentle placement of a dolphin calf in their group. It was something to behold.

After our days in the dolphin area we headed to the Abacos, the beautiful islands on the northeast side of the sandbank. Patxi taught the kids to scuba dive as we explored various reefs along the way. Patxi and Bill were both to become supporters of my work and Bill became a friend over the years. Bill's kids, now grown, still vividly remember their experiences, which were special to them, and probably to the dolphins as well.

May 28 heralded the start of our regular season and once again Rosemole was the first dolphin to show up at our boat. Already my favorite dolphin, I was excited to see her but thought it both amazing and strange that she arrived first, never having seen the *Wren of Aln* before. We began to call the first encounter of the season the "reunion." The level of excitement both with the dolphins and us was tangible and electric. I have no doubt they recognized us in the water. Perhaps they labeled us as the seasonal migrants that return to mark the start of the summer season. Either way, "joyous" is probably the word I would use to describe it. And even though I am committed to studying and understanding the dolphins scientifically, I have no problem also feeling like they are friends, of another species, but clearly aware, with feelings and memories, and this was a reunion of friends. I have often wondered how they recognize us. Is it the constant presence of my video camera? Is it the way I swim, or the funny sounds I make through my snorkel when greeting them? I imagine it could be all those things, as dolphins are exquisitely adept in their own long-term relationships that would require such recognition.

By now the pace of the summer season was deeply embedded in me. Some days we anchored our vessel in our main study area. These days were often lazy, cloud-drifting days, where dolphins chased ballyhoo and needlefish under circling frigate birds. For hours and hours we watched them, 360 degrees around the boat. We learned quickly that it was impossible to observe the dolphins underwater when they were chasing schooling fish. Sometimes we slipped in the water and got lucky as a young dolphin swam by inverted just under the surface, eyeing a doomed needlefish above. Ballyhoo and needlefish frantically tried to flee the dolphins while the frigate birds overhead waited for any escapees. I watched from a distance as dolphins chased fish and frolicked for their lunch. It was, and still is, one of the most peaceful and satisfying sights for me, knowing that the dolphins are making their daily living with such ease and without constant human disturbance.

The shallow sandbanks of the Bahamas are really an amazing place if you are a dolphin. There are miles of submerged island, which provides a safe place to rest and socialize since sighting a large predator on a white sandy bottom is easy. The sandbank, richly endowed with diverse fish life under the sand, provided snacks throughout the day, dainty treats of razorfish, snakefish, eels, and flounder, the last a favorite of the inexperienced calves.

During the daytime I observed a variety of different feeding techniques on the shallow sand. I saw that flounders, resting on the surface of the sand, were easy targets for the younger dolphins as they scanned along the bottom, barely stopping after their catch. Snakefish, hidden under the sand, darted in various directions after the dolphins buzzed them from the surface and began their chase. Razorfish and eels burrowed even deeper and were hunted as the dolphins dug deeply into the sand itself, eventually snagging their buried fish. Sometimes the dolphins ran "transects," skimming the surface of the sand in patterned grids while they searched for fish. Leisurely working parallel to the ridges and facing the current, groups of four or more synchronized their tail slapping on the bottom as they scanned and cruised. Sometime they scanned from the surface in a contorted posture, similar to the *S* posture, directly down with no audible sounds, and then quickly zoomed down to the bottom to snag a fish.

Bordering the shallow edges are the pristine coral reefs that nurture life in many forms. Reef fish, octopus, turtles, and sharks all benefit from the nutrients of the deep scattering layer (DSL) and the tides as they ebb and flow on and off the sandbank every six hours. Grass beds to the east provide a wealth of nutrients that flow off the bank, not to mention a safe haven for juvenile fish and invertebrates. The rhythm of this flow is at the same time both subtle and obvious. Edges of the tide line, green from the sea grass bed nutrients, flow westward to the deep at outgoing tide. Clear cobalt water from the adjacent Gulf Stream forms shades of blue that return back to the shallows as the tide flows eastward back on to the sandbank. Most animals inhabiting the sandbank adjust to this ebb and flow in their lives. Like these creatures, we humans adjust our daily routine to incorporate the same rhythm and flow, although we are perhaps not as acutely aware of how it drives us.

The geology of the sandbank is basically limestone. There are no underwater volcanic sources of land, no lava. Only the precipitating oolite, calcium carbonate that chemically falls out of the warm seawater and descends like rain out of a cloud. It is this cement building material that the reef uses for its structure. It is also the process that will most likely be altered by the rapid acidification of the oceans, a process so potentially destructive to all shell-forming organisms, and virtually all ocean life, that I tear up in embarrassment for my species, and find words of apology to the ocean life that somehow don't seem adequate—especially in this fine moment, when the rhythm of life seems easily in tune with itself.

I awoke out of my reverie as suddenly a familiar mother-calf group of dolphins, surrounded by protective male adults, appeared. It was a fine June day and Diamond, the offspring of Luna, and other young calves swam along in formation with their mothers. Diamond suckled for five-second bouts as mother and calf glided along. A large nurse shark cruised by on the bottom with a remora on his left side. The spotted dolphins disappeared briefly but then rematerialized a few minutes later with four old adults who circled and watched the young dolphins. There was no great distress among the dolphins while this harmless shark was around. Nor was there much noise to shake me out of my dream state and I continued to watch as both the dolphins and shark drifted off. I observed the rear guard of adult males escort the young dolphins away from the shark and wondered how they communicated this simple movement.

Dolphins are, no doubt, masters of their acoustic world, but they are often silent, which seems counterintuitive in some circumstances. Occasionally the shallow sand from the bottom is violently kicked up from storms or distant swells and fills the water column with fine grain particles, giving it a milky appearance. And the spotted dolphins are silent. You would think in murky water they would use their echolocation even more, but dolphins, it turns out, are good listeners and they do a lot of it. Our nervous system is designed to filter irrelevant information out. Over time, and with experience, we learn what sounds are significant and which ones are background noise. I have often watched dolphins at twilight in the deeper seventy-foot water, swimming along quietly, before they hear a cue on the bottom that suddenly orients and directs their now audible and active echolocation that way. In moments of quiet it is possible the dolphins are producing ultrasonic signals that our ears cannot perceive. But to date, I have not recorded any ultrasonic signals produced in these situations (meaning above human hearing at twenty kilohertz) with high-frequency sound sampling equipment.

Sometimes the dolphins led us to new places. In mid-July as we ran down the edge of the sandbank, Flying A and her southern friends appeared at our bow for a ride. The southern dolphins were more elusive and harder to find so we were excited to see them. As we watched their hydrodynamic act on the bow to my horror I suddenly saw the shallow eight-foot reef loaming up, a frightening sight for our boat, which only has a six-foot clearance. The dolphins had led us into an area we call the Dry Bar, named for waves that break on the shallowest part of the reef. Well-marked on the charts as "rocks," we had avoided this area over the years, but now the dolphins were leading us there. Over the centuries many a ship has met its death on the Dry Bar, and before we became an additional one, we veered back toward the deep water. In the days that followed, and with good sunlight, we explored the Dry Bar, snorkeling the many cannons and wrecks. We eventually met Mel Fischer's marine archeological team from Key West, Florida, during their excavation of what appeared to be a Columbus-era vessel, conquistador hats and everything. History was everywhere in the Bahamas and most of it was underwater.

In early summer I received a surprise phone call from the UNEXSO dive shop in Freeport. They had just received a young spotted dolphin that had been stranded in the Abacos, for some unknown reason, and they wanted to release her. Had I known more about the challenges of dolphin releases, I might not have agreed. But it seemed like an ideal opportunity to see if we could put this young dolphin back in the wild where she belonged. She was, in fact, a pantropical spotted dolphin, *Stenella attenuata*. We were set to release her at least close to home. At that time I had the sense that she might be accepted into our local group, but as I write this book I reflect on the release of a young Atlantic spotted dolphin, Cutter, I had been involved with in 2009. Cutter was stranded in the Florida Keys, rehabilitated over three months, and released in a large group of spotted dolphins off Key West. Although radio-tagged,

Cutter was never tracked by radio signal, or seen with his visual freeze-brand on his dorsal fin after the first day of his release back into the wild. Cutter had the most ideal of circumstances and chances to reintegrate into a group, but the ending to his story is still a mystery. But, regardless, we had agreed to help this young pantropical dolphin and we rapidly made plans.

The plan was relatively simple. Mike Schultz from UNEXSO, along with John Englander and others, would prepare the dolphin and bring her up to our study site to release on July 22. They were supposed to have freeze-branded her dorsal fin so we would have a chance to identify her again and determine if the release was successful. The day came and the team headed north from the island to meet us in their small twenty-foot boat with the young dolphin in a small child's pool on board. On our vessel, the *Wren of Aln,* we searched and searched for our spotted dolphins. Eventually we met up with the small boat and I hopped aboard. Our research vessel left to continue trying to find a group of dolphins. Soon our task became a life-threatening nightmare. A monster squall set in upon us, and as a small boat in a large sea we were about to capsize. Luckily, one of the local treasure hunting boats, the *Beacon,* was nearby and we radioed our desperate situation to see if they could help us haul the pool and dolphin to their deck while the storm passed. I have always tried to make my graduate students understand how dynamic the weather is in our field site, and how important it is to monitor and take precautions so as not to get caught as we now were. Meanwhile, the *Wren of Aln* had just found a large group of thirty spotted dolphins and was headed our way. With enthusiastic cheering on the radio, the crew and passengers encouraged the bow-riding dolphins to stay with the *Wren.* As the *Wren* approached, and as the squall veered away, we decided it was now or never and we slipped the young dolphin overboard. All our plans of temporarily housing the young dolphin in the water to expose her to our study group slowly went out the window. This was her shot at freedom. The group of spotted dolphins leaped out of the water and escorted her away from the boat in what appeared to be a welcome, dolphin style. Unfortunately, the freeze-brand had not been done by UNEXSO for reasons unknown, so the young dolphin was off and about, only recognizable by her species marks. We named her Splash after her dramatic entry into the water. In the next two weeks we observed Splash in the company of bottlenose dolphins three times. Then Rosemole came by one day to solicit a game of sargassum play, a favorite interspecies pastime. As a large pregnant bottlenose dolphin swam by with two small calves I saw that one looked like Splash. Rosemole positioned herself in front of me and "wagged" the seaweed on her tail, but I ignored her, anxious to watch the bottlenose dolphin to determine if Splash was there. I felt bad that I ignored Rosemole's overt invitation and I was glad she came back later after I had snubbed her. In the end it was still unclear whether I saw Splash with the bottlenose dolphin and we never saw this released dolphin again.

I have thought about releasing dolphins back into the wild for many years and now have more definite opinions than I did during the release of Splash. While we wanted to give this young dolphin a chance at freedom, we were clearly not releasing her with her species. Although it is not unknown for interspecies adoptions to occur, both here and in other parts of the world, it is not the norm. It's possible that Splash was rejected from the spotted group and was now with the local bottlenose dolphins.

I am still glad we tried. But even with the best of scenarios, like Cutter's release, it was still a difficult task. Some argue that a chance at freedom is better than captivity any day. Others argue that such an animal is not releasable. I am somewhere in between and I believe that if they are candidates for release, that is healthy, socialized, and from a known area, they should be given a chance at freedom. Releasing dolphins in a group is probably better than releasing single animals, since they are a social species and need both protection and emotional support in their daily lives.

Dolphins in the wild have their challenges, both with predators and their own health. One early morning as we watched the dolphins feed lazily under a frigate bird, a lone spotted dolphin approached our boat. He was disoriented and the underwater video captured his odd echolocation clicks. His blowhole area was "dented"—perhaps a deformity but more likely the "peanut" head described for emaciated dolphins when they lose fat from the forward part of their head. My feeling of discomfort escalated in the water because as he approached us he didn't see us, and passed too close for comfort. Later in the summer we saw this dolphin, now named Denty, and he was clearly emaciated from his inability to hunt using his sound. His head was even more indented, as if he were burning his reserve blubber. We tried to think about ways we could help him, with antibiotics and food, but decided to not interfere with natural processes. In the end we let nature take its course. Only in instances of human impact, such as a fishing line wrapped around the tail of a dolphin or nets, have we intervened. Nature is one thing, human-derived problems are another, and like the researchers that unwrapped right whales and humpback whales from crab pot lines, we, too, feel the need to help when our species has caused the problem. By late August Denty was still trying to catch fish but even a needlefish on the surface was an exhausting chase. As he swam by the video camera, I saw new wounds on his left side in addition to two remoras, opportunistic hitchhikers on wounded dolphins. Denty was not seen again.

Late summer took on its own pace: hot days, warm water, and hurricane threats. It is routine on the boat to listen to our VHF radio for weather, a bit of a ritual. I admit I am something of a weather junkie and enjoy the power and dynamics of storms both at sea and on land. By late August we were monitoring multiple storms. Tropical Storm Felix was developing off the Canary Islands; Hurricane Gabrielle was approaching the Virgin Islands then quickly moved past Puerto Rico toward Bermuda, sending ten-foot seas our way. On September 14 Hurricane Hugo was building near the Leeward Islands to the southeast of us. By September 16 Hugo was too close for comfort, so we shortened our trip and ran back to Florida with our passengers. Hugo devastated South Carolina and we escaped once again from a major hit in Florida. Hurricanes are a force of nature and often unpredictable, creating a stressful addition to our fieldwork.

Every year I tried to get out in my field site off-season for a more accurate picture of the dolphins' lives, but it was never easy, usually because of weather. This fall I had an October trip scheduled to the Bahamas. While moving my gear from my home to the boat (at that time my office was the second bedroom in my rented apartment), I was burglarized. Fortuitously, one week before I had purchased renter's insurance, and I knew I could get reimbursed for some of my losses, but it was my underwater video gear, and a VCR for the boat, that had been taken—my work equipment! Panicked I called my friend Patxi, who had various sponsorships from Sony in his diving career, and without a second thought he sent some equipment up from Miami for my trip. He barely remembers this generous gesture but it saved the trip. After this uneventful research trip, which included a rat from shore chewing my dive mask to pieces on deck, and a scary Gulf Stream crossing with seventeen-foot seas (the first cold front of the season), I returned to find that the police had found my video housing unit at a pawnshop, recognizable because of the painted dolphin on the cover. So much for working in the winter months. I happily settled in to my new life in Florida and spent my days processing slides, analyzing data, and planning for the next field season.

1990—Films and Fins

Soon the field season began again and the Wild Dolphin Project had its first film project with the BBC. Renowned naturalist David Attenborough was to narrate a series called *The Trials of Life.* In the middle of April I boarded a ship called the *Seaborne,* commandeered by the BBC's film crew. Although a bit early for my normal field season, we set out for the Bahamas to film spotted dolphin behavior. The vessel itself was a bit like a houseboat and looking around I could see no life preservers. The owners were more than hospitable, letting me dethaw occasionally in their bathtub after working in the brisk northeast winds all day. (Yes, it does get cold in the Bahamas.) As I was to learn throughout my career, film crews need to be busy, and this film crew was no exception. Upon arriving on the site, Mike deGruy, a well-known and talented natural history cameraman, became immediately anxious and wanted to run around in a small boat trying to find the dolphins. My experiences over the last five years had made me realize that it was best to anchor and wait for the dolphins to come in on their own time, which they did on a regular basis. Although giving this advice was my job as the dolphin expert, it went unheeded. For hours and hours the film crew buzzed around in a small boat trying all sorts of tricks, including starting the engine to attract the dolphins. Exasperated, but finally open to my suggestion, Mike said, "Well, let's try it your way." So we turned off the engines, we turned off the generators, and we waited. Within twenty minutes a group of dolphins headed to the boat and the film crew got busy in the water. That was one of many times I felt vindicated by my real knowledge in the field and how we worked differently, and cooperatively, with the dolphins.

By now I could also navigate visually on the sandbank, at least on a nice sunny day. With no land in sight, my eyes followed the underwater depth markers, white sand shoals, coral reefs, and the deepwater edge to navigate up to the dolphin area. One such journey was particularly striking, as the captain of their vessel had extremely bad eyesight, and I led him to our anchor spot off a small wreck, the dolphin wreck, visually. I am sure the local Bahamian fishermen do the same thing with their intimate knowledge of the reefs and sandbanks. Obviously the dolphins can navigate with their senses, and although I could navigate on my boat, I still felt like a rookie in the dolphins' world.

Later in the trip David Attenborough arrived and the film crew concentrated on getting daytime shots of him in the water with the dolphins, and his thoughts and reflections on camera. One day as we watched a large group of dolphins swim by parallel to the boat at a distance, David said to me, "Why don't you go bring them back?" Maybe he was kidding, but this was something that I had actually done before. I had learned when I was alone in the group of dolphins I could use their own signals, which included sharp but subtle head turns, to direct the group. I had observed that this is how they themselves changed direction as a group. Although I wasn't quite clear on really who had the authority to initiate such turns, it occasionally worked. So I calmly swam out to the group, got in the middle of them, and gently nudged my head 180 degrees back toward the boat. As the dolphins and I approached the boat, David slipped into the water. For the moment I felt quite the expert, but I learned later that these signals didn't always work. I believe that such moves were based on which individuals had the authority to make such decisions (either by age or rank). Clearly humans were low on the list, but I was lucky that day.

While sometimes I became part of the dolphin group and influenced their direction of travel, other times I was put in my place. During one incredible encounter a group of dolphins had allowed me to join their pod. I decided to experiment by moving into the middle of them. While they kept their eye on me, the dolphins guided me with simple head nods and direction changes ever so gently back in my place, which was apparently in the rear of the group. Gentle persuasion, it seemed, was the mode and I suspected this is how they trained their young to keep a certain location in the group for protection. In my case I was a pathetic human and probably needed the security of this position. I wasn't really trying to control the group, but I was trying to learn what signals they used to communicate in the group. A few weeks later my own research vessel, the *Wren of Aln,* arrived on site. I was never so happy to see my boat. I immediately swam over to gorge myself on the fruits and vegetables, supplies that the English film crew didn't feel were necessary. By this time I was quite tired of Yorkshire pudding and roast beef.

In those early years we were often the only boat up in the dolphin area. Occasional dive boats came up and tried to get some time with the dolphins in between their dives on the nearby reefs. But we were focused solely on the dolphins and stayed in the dolphin area, so we often had the sandbank to ourselves. As the years passed, and the boat activity increased, it was no longer easy to be the only boat in the dolphin area. Annoyingly, dive boats circled us and tried to steal the dolphins away from us with their lure of a bow ride. Eventually we agreed upon some boat etiquette, which meant that we kept a half mile apart out of respect for anchored or operating boats, but initially it was a pretty rude scene. Boats charged each other and threatened to ram the other as a display of territory. I laughed when I thought of the similarities between that behavior and the dolphins' head-to-head aggression; just another example of mammalian territorial displays. I wondered if humans would ever evolve above such behavior.

In May the water temperature already registered eighty-three degrees. Typically the water was barely eighty degrees by June. Did it signal an early hurricane season? Throughout the summer we settled into our normal routine: we anchored the boat, had three or four encounters through the day, which was average, and got up and did it again. Weather was always an issue, so if we had to move we sheltered in the lee of the nearby reef to break the swell, which allowed us a good night's sleep.

I tried very hard in the early years to make the relationship with the dolphins a priority and to keep our relationship in good order. I knew that we had to win the dolphins' trust and behave ourselves in the water if we wanted to observe their natural behavior. I took this extremely seriously in my first years: I just anchored the boat in one location where I knew the dolphins swam by. And I just waited. I waited for them to be curious and swim close to the boat. When they came by we slipped in the water and tried to get them used to our presence benignly. We were armed with a camera and snapped a few shots of their spot patterns. We swam slowly, kept our arms behind our backs if we weren't holding a camera, and tried not to chase them. Some dolphins were more curious than others. Some dolphins were wary and kept their distance. Eventually we began to recognize certain individuals. Sometimes they came by when we humans swam laps or jumped off the side of the boat for fun. I think they were curious about us, and wanted to see "normal" human behavior. Sometimes they spy-hopped near the boat, and watched us on the deck of our boat where we ate and laughed. We became a part of their environment without being invasive, although we made noise just with our generators running. By now I realized that this was the most important investment I had made in the relationship with these wild dolphins and it would pay off for decades.

This season all sorts of people came and went. I was always amazed by the groups of people or boats that showed up in the dolphin area. Especially amazing was that they came with very large goals, and sometimes not even knowing what species of dolphin they were looking for, nor much about their behavior. On June 11 a Cousteau film team showed up. Of course Jacques Cousteau drove my interest to explore the ocean as a young girl, so I was familiar with their iridescent swim skins that flashed in the bright sun as they zoomed by aboard their small inflatable in search of dolphins. Later that week, after finding it difficult to get time with the dolphins, one of the crew came over to our vessel to chat. During our talk on the back of our deck the dolphins showed up three times, lazily sauntering in toward our stern, during which time we slipped in the water and recorded some behavior on our video and slipped out. With no luck in filming the dolphins the Cousteau crew listened as we explained that these dolphins were used to anchored vessels and that they often took the initiative. The usual human approach to wildlife interaction had always been too aggressive for my taste. Of course time is money for film crews, but some of the most professional photographers that have been out with me over the years knew how to pace themselves and blend in. To me, it was often a sign of professionalism. Well-meaning intentions turned into inappropriate etiquette for the species involved, and that was often the case with the dolphins. For our work, appropriate etiquette made all the difference.

In mid-June a research group, from a local human therapy clinic, came out to the field site with an armada of boats, and at their request I went over to chat with them. During our discussion I realized that they were here to "crack the code" of dolphin language, in ten days, by putting a speaker in the water. After further discussion I realized they didn't even know what species of dolphin were here. I may not be an expert on linguistics but I do know if you want to understand another species you might want to know a bit about them such as which species they are, who they are as individuals, and the society's rules. Of course that was my framework, which really was more akin to an anthropologist than perhaps a biologist, but in my mind this was a unique situation and opportunity. And I wanted in with the locals. But many people have a New Age perspective that inappropriately elevates dolphins to gurus whose lives in the ocean rotate around waiting to interact with humans. Nothing could be further from the truth. Although I am the first one to acknowledge all the incredible interactions with these wild dolphins, I realize first and foremost we need to let dolphins be dolphins in the wild.

But then again the attitude that dolphins needed to be like humans before we can ascribe intelligence to them wasn't so surprising. Scientists, in the early years, tried to teach primates English, even though chimpanzees don't have the vocal structures to make such sounds. Later they tried to teach dolphins to reproduce English words, and as you can imagine dolphins are unable to produce human sounds, let alone words. This human-centered attitude is analogous to the mistaken belief that the sun circled around the earth. Today, even in the most progressive of sciences, our human and primate-based biases remain. Their ten-day project to crack the code failed and was a disappointment, but perhaps a more realistic group of humans returned to land.

Because of our intimate relationship with many individual dolphins, one of the hardest things to see was a sick dolphin. Dolphins are mammals susceptible to all the things humans are, including bacteria, viruses, fungi, and so on. For dolphins in the wild it did seem like a slow, somewhat torturous death. Hugh was a young dolphin we observed that summer from our Zodiac. Disoriented and weak, he rubbed on the bottom and tried to dislodge his remora. Even though I had observed a mother trying to remove a remora from her calf, it only caused the remora to move around on the body and clasp tighter, leaving a blue gray scar resembling the bottom of a tennis shoe. Hugh dug in the sand with his rostrum to rout out fish, but nothing appeared. Only once did I hear him use his echolocation while chasing a horse-eyed jack. Even his eyes were foggy and lifeless. His whistles were scratchy and harsh, unlike normal signature whistles. And, like Denty, he was alone. This is typical of sick dolphins on the sandbanks. If they heal they rejoin the group, but sick dolphins are likely targets for predators and are quickly recycled back into the system. We saw Hugh one more time in September; he approached us with an inverted tail slap, briefly swam by, and produced no vocalizations. He looked emaciated and had a small dent on the side of his head. He looked worse than at our last sighting and we never saw him again.

We often had small land birds, including various species of warblers and purple martins, land on our boat at sea, which we rehabilitated and brought back to the island when possible. After storms, especially if the wind blew from Florida to the west, herons, owls, and purple gallinules landed for a brief rest on our boat. We were, after all, the only island mass around. We saw the magic of large manta rays swimming against the white sand bottom and sailfish leaping out of the deep purple water, a reminder of our proximity to the deepwater edge where large creatures make their way in the world.

I had many dark days in the field as well. There were some days when I just didn't know if I could go on with the challenges of the project. Although it was a joy in countless ways, the responsibility often seemed overwhelming, financially and personally. Part of me realized that I might never have a normal life, and it wasn't that I even wanted one, but I did miss normal things of land life. I was bicoastal for years, living in San Francisco in the winter for graduate research, and doing fieldwork in the Bahamas in the summer. Friendships and relationships were strained and hard to solidify. My cats had seasonal company. I went back to a dark, cold basement apartment in San Francisco after being in the field for four months, which was depressing. My move to Florida in 1989 after finishing graduate school solved part of the challenge, but not all.

It was also a lonely journey much of the time, partly because I felt the responsibility of the work so personally. I knew this was a unique opportunity, to tell the story of a normal, healthy dolphin society underwater, and intimately, for the first time. I wanted to do it respectfully, carefully, and creatively. I had been alone much of my life, having lost my parents early. It was in these times that my long-term friendships with Linda, Chris, and other friends that helped me shape the work and the project were invaluable. I knew my methods in the field were a bit unorthodox and not by the book, but I also knew that the only way to work underwater and give testimony of the dolphins was to make sure they were equal participants. In many ways I didn't care what more conservative scientists thought about my methods. I was young, naïve, and quite stubborn, and I knew I was breaking new ground with rules yet unwritten. I trusted that the dolphins would guide me, or reprimand me in some cases, should I go out of bounds. I have always tried to work benignly with the dolphins, using only hands-off methods to track them through photo-identification, trying to respect their communication signals and system in the water, and to follow their rules and etiquette, as much as a terrestrial species can in their alien watery environment. Although the dolphins are inherently tolerant of our mistakes or missed cues, we take their signals seriously and reiterate to the dolphins that we are aware of our mistakes and are trying to rectify them and act appropriately.

After summer field seasons I escaped into the mountains of Maine to refresh my soul and remind me of my own land-based roots. In the woods I felt the pace of the earth moving, its decay and change, both differently paced and strangely visible unlike much of the water world I emerged from. Although the seasons and pace of my own fieldwork were now ingrained in me, the perception of them came through different senses: the smell of salt air, the violent heaving of the seas, the limited senses while working underwater. In my dreams I heard the dolphins snickering, wondering why we terrestrial aliens flounder so in the sea, seeking to master another medium that will always be foreign to us. "How bizarre," they would squeak.

Early morning encounters with the dolphins are my favorite memories.

I am usually up on the boat before sunrise. I need my coffee to wake up and it is very harsh to jump in the water asleep. Sometimes the dolphins were at the bow, or near the boat, when I walked outside with my cup of coffee. These were special moments: it was quiet, no one else was up that early, and the dolphins just hung by the boat as they beckoned me to slip in the water. Sometimes mothers stationed their calves on the bow as the strong currents flowed past, teaching their inexperienced calves to position themselves on the bow in this situation without the danger of a fast-moving boat. Early morning interactions with the dolphins were usually slow and mellow as if the dolphins had just come back from the deepwater edge of all-night feeding and were resting.

I sometimes slipped in the water and had my own private sunrise encounter. In my later years, after I had seen one too many large tiger shark and fought too many strong currents, I was a bit more hesitant to get in the water alone without at least one person on deck in case of problems. Sometimes I just dangled my feet over the bow where the dolphins hung and I watched them. Sometimes they spy-hopped at the surface, or blew sounds through their blowholes at the surface. It was truly a crossing of boundaries, the dolphins in the water and I above the water, a place where we could meet. What would it really be like to talk to another species, to understand? Perhaps it was more of a communion than a communication, but these moments were special nevertheless.

I remember waking up one morning when visibility looked murky and the current was strong yet here was a mother and young calf hanging on the bow. One of our passengers, an underwater photographer, was up early on the deck. I invited him to go in with me, but he refused. To him it wasn't worth it without the light for photography. How sad to miss such an experience, because not everything in life should be limited to our photodocumentation. The image is still as clear in my mind as a ten-pixel image but he missed it.

I usually approached the issue of etiquette and rules in the water with caution. Sometimes I tried something and if I noticed a bad response from the dolphins, I tried to not break the rule again. Sometimes the dolphins taught me where to swim, when to dive down and turn inverted, and when not to. It was always a learning experience. It took me a while to really understand the etiquette of the dolphins. One time while following a male dolphin as he dragged sargassum on his tail I reached out to pull it off his fluke. At the time I didn't know this was inappropriate behavior. The rules were about the game. If the dolphins want you to have the sargassum they drop it in front of you. This male didn't like me grabbing his sargassum and he blew a large bubble and left. I never made that mistake again. Although these were all personal moments, beyond the recording of their behavior, every one of these insights has guided my behavior in the water over the years and allowed me to observe closely, and often unnoticed, intimate details about the dolphins' lives.

By late July we had some major hurricanes brewing. A tropical depression was just north of us, Hurricane Bertha was heading our way from the southeast, Tropical Storm Caesar was forming off Africa, and tropical waves were whipping up in the southern Caribbean. Large swells from the northeast pounded our study site. Yes, it was hurricane season in full swing. We managed to stay in our study site through August, but as mid-September approached Hurricane Isadore formed off the Azores, and a week later we headed home to Florida.

The Gulf Stream is an amazing place housing many deepwater species of dolphins and whales. Pilot whales, a member of the dolphin family, are regular migrants in the Gulf Stream. On this blustery September day we watched as this family group traveled north, riding the current of the Gulf Stream in parade formation. There are many pelagic species in the Gulf Stream including spinner dolphins, pantropical dolphins, and large whales. We saw them in the summer months, but also in the winter months when occasionally we ventured east to the Abacos for a winter trip. Today the whales moved north, as the Gulf Stream does, northward to fishing grounds and other opportunities. Some species move and some species like the spotted dolphins in the Bahamas are resident, finding enough local food to eat. Myself, I am both, moving between the human world and the dolphin world, finding different types of food that feed my soul.

1991—Rosemole Gives Birth

It was May 1 and the water temperature was already eighty-two degrees, four degrees higher than normal—unheard of, for this time of the year. Typically in the spring the water is around seventy-eight to seventy-nine degrees, and it gradually increases throughout the summer. By August it is typically eighty-eight degrees, almost too hot to swim. But in May I don a short wet suit, since my Minnesota blood has thinned, and even though the water is warmer than usual the northeast winds keep the air brisk.

Early May proceeded as other Mays, with a nip in the air and new calves around. Nippy, now the mother of a two-year-old named Nassau, darted around our boat. Old friends Little Gash and Mugsy were around, but more and more they had a serious side and spent time with adult males in courtship activities. Rosemole gave birth to her first calf, Rosebud, this year, which altered the regularity of Rosemole's visits to our boat. Rosemole was now associating with other females with calves as is typical for new mothers. I had seen over and over how important it was for a mother to discipline and calm down a rambunctious calf. Now the light touch of Rosemole's pectoral flipper to Rosebud's body served to calm her down and Rosebud's previously loud excitement vocalizations ceased. They swam away together, keeping in constant touch as they glided through the water. Rosebud followed her mother, keeping light contact with her fluke. Rosemole tail-swiped Rosebud's head gently and Rosebud slowly chased her mother while they kept in physical contact, even as Rosemole rolled around. Then they switched positions and Rosebud was ahead of Rosemole, gently stroking her head with her fluke. As

Rosemole positioned her rostrum to Rosebud's genital area with her mouth slightly open I could see Rosebud's delight in the sensation. Was Rosebud soliciting her mom for stimulation? Sometimes a mother buzzes her calf this way to discipline her, and at other times buzzing is used for play and social interaction. Sometimes it is hard to determine who initiates a behavior and, as some new models of behavior analysis suggest, social behavior may be dynamic and interactive in real time, rather than hierarchic and sequential. Rosemole's mother had probably disciplined her in the same way and now she was using her knowledge with her own offspring. Despite her new responsibilities, Rosemole came around the boat for an occasional game of seaweed, but as the needs of motherhood grew, my time with Rosemole lessened. It is the way it should be in the wild and despite my loss of playtime with Rosemole I was ecstatic to see the development of her skills as a mother.

Now that the juvenile females I met in 1985 were mature and giving birth, the patterns of courtship and the struggles of motherhood were becoming clearer and clearer. Males continued to surprise me with their unique repertoire of behaviors during courtship. Males chased groups of females with their necks out of the water, straining to stay at the surface. This "rooster neck" chase created a wake in the water and it suggested an intensification of the normal *S* posture signal that a male used when courting a female underwater. Prosodic features of communication, such as timing, intensity, and spacing, are used by other species besides humans: wolves synchronize their howls, chimpanzees coordinate their fights, and dolphins showed synchrony and rhythm in their acoustic behavior and the intensification and coordination of their body postures.

If all went well Rosemole would be a successful mother and have a calf every three to four years after she weaned Rosebud. Calves stay with their mothers for two to three years or until she becomes pregnant again. Pair One, a new mother in 1990, had just lost a calf, sending her rapidly into estrus and causing Punchy to chase her intensely for a mating opportunity. Meanwhile, Punchy's friend, Big Wave, held Pair One's older calf, Poindexter, between his pectoral flippers, as Punchy mated with his now available mother. Mugsy, after the loss of her first calf, also engaged in mating behavior during postpartum. Apparently postpartum mating was normal, calf or no calf.

Sometimes males overstep their bounds and females react quickly. As two males approached a pregnant and lactating Nippy, she tail-whacked both males on the head as they tried to approach her in a typical mating attempt. Another female, Barb, blew a large bubble at the surface as two males tried to copulate with her. The females in this group were having nothing to do with their interested suitors and amply communicated this feeling.

Meanwhile Rosemole's and Mugsy's friend Little Gash was still not pregnant, but was eagerly and seriously engaged in courtship behavior with the males. Little Gash was often the only female among males. Stubby and Knuckles, her primary suitors, flanked and rubbed her. As Knuckles moved in upside down to mate with her, Stubby dashed in to mate with her as well. Interspersed with their courtship activities, Stubby escorted Little Gash down to the bottom to forage, while Knuckles meandered above on the surface. This is a monopolization strategy of the males to go along with the activity of the female. Little Gash, not being one to accept an undesirable suitor, tail-slapped one of the males away as she displayed some aggressive head-to-heads, open mouths, and squawked. Usually a calm affair, Stubby and Little Gash's mating was paralleled by two juveniles mimicking their belly-to-belly activities, typical juvenile behavior. Dolphins have many strategies to learn, which include watching, practicing, and participating.

Teaching young dolphins is a cross-gender affair. Adult males use the same methods of discipline and expressing displeasure as female dolphins. Latitude, a rambunctious calf, made his signature whistle and tail-slapped wildly on the surface as Big Wave calmly went over and touched Latitude gently to calm him down. Meanwhile, behind Latitude was his mom Luna. She approached her calf inverted (usually a sign of extreme displeasure in mothers and impending doom for the calf). She chased and swam after Latitude rapidly. Later, Havana and Latitude (friends in crime) overlapped signature whistles excitedly when Hedley, Havana's mom, came by us and whisked them away. The calves wanted to play, but the adults didn't want them to be distracted. Mothers usually had the upper hand.

June came and went, full of dynamically driven waterspouts and rain-squalls, typical during this weather transition month. Nippy's new female calf, Nassau, became a regular visitor and Nippy often brought her by the boat. In early July we entered the water only to see little Nassau with a large, but shallow, shark bite. She was also missing the tip of her dorsal fin. Two years of age is a vulnerable time and Nassau had a close call. It's the dolphins we never see again who are probably not as lucky. Nassau rubbed on the sand bottom and tried to dislodge the dead skin from her wounded side. Nippy slowly foraged nearby and repeatedly whistled to her calf, keeping in acoustic contact. In late July we saw Nippy once again, but where was Nassau? She was young enough to still be with her mother, but she was nowhere in sight. Nippy foraged in the sand, while she carefully avoided the annoying little tunny that darted in and out and tried to steal the small fish morsels left behind. Only in mid-August did we see Nassau again, this time with Nippy. It was a mystery why this two-year-old dolphin wasn't with her mother during part of the summer. Although we had seen adults and juveniles remain solitary during an injury, a young calf's isolation was not safe. Perhaps she was able to catch enough fish to sustain herself, or perhaps she had the company of other dolphins. We will never know, but at least she survived.

This summer, my old mentor Ken Norris visited us on our boat. Joined by *National Geographic* photographer Flip Nicklin, Ken was writing an article entitled "Dolphins in Crisis." Traveling around the world, Ken was bearing witness to not only the variety of dolphin species, but also the variety of impacts humans had on the habitat of cetaceans worldwide. Flip joined us for six weeks to get some working shots and document our research. He photographed a full-body image of me in the water with my video gear, and I jokingly told my relatives after the article was published in 1992 that I was a centerfold with the crease of the magazine cutting through my now-globally exposed thighs.

Ken in many ways is the father of dolphin research. As a student in San Francisco, I eagerly drove down to Santa Cruz to listen to Ken lecture in his marine mammal class. Originally a herpetologist, Ken is an old-school naturalist and he inspired many a graduate student. Fellow student Jan Ostman went off to Hawaii to work with spinner dolphins, Michael Poole to Tahiti to also work with spinner dolphins, and I to the Bahamas to work with spotted dolphins. Ken was adamant about getting a look at their underwater world, and he had built a boat with a sturdy plastic tube projecting below, large enough for a researcher in Hawaii to sit in and watch spinner dolphins. Known as the seasick machine, it provided mostly a behind-the-group view as I was told later by his students. So it was not a surprise how shocked Ken was when he joined us in the clear blue waters of the Bahamas and got a look at spotted dolphin behavior underwater. Having never seen this species, he remarked how different they were from spinner dolphins. "You would never see a spinner dolphin away from the group," he remarked as we watched a rambunctious juvenile spotted dolphin play around the boat.

Later that night, drifting along the deepwater edge with glittering stars overhead, Ken told us the story of spinner dolphins in Hawaii. He talked about the volcanic islands and the habitat this provided, about the rhythm of the spinner's day, offshore in deep water at night to feed, sheltering in the shallow bays during the day to rest. He was a storyteller, and a good one. His keen eye for natural history and his strong sense of the need to check back with nature after data collection were qualities I have always valued and tried to incorporate into my own work. I often wished I could share with my graduate students the wonders of fieldwork as well as Ken did. Today, students come to collect data, forgetting to look up at the stars and think about the pulse of the animals they are studying and their environment. I think it is a loss for the world, to reduce animals to data, instead of telling their story. As my long-term mentor and colleague, Scott McVay, once told me during my early years trying to describe the underwater behavior of the dolphins, "It may not be as much about describing the behavior, as understanding the process of communication." I think these words hold true more now than ever, and that our work propels us to understand the laws of nature, as much as the details. Scientists can, and should, do more to study the laws and dynamic of the communication process.

One of the traditional ways to study animal communication is to playback vocalizations and to observe what behavior occurs and how the animals react. This was our third and final year with our chartered catamaran the *Wren of Aln*. It was also the first and last year I tried some playback experiments with the dolphins. I have never been a big believer in playback experiments in the water because most of the time we are unable to observe the real reaction of the animals. Over the years I had collected the signature whistles of many of our regular dolphins and since dolphins use their whistles to identify themselves as well as contact other dolphins, I thought it might be worth a try. To playback these sounds my assistant operated the computer inside and I remained in the water to observe while we played back their whistles through an underwater speaker connected to the computer. I played Katy's whistle first since she had been here in the morning. As dolphins arrived, and as I identified them in the water, I instructed my assistant to play their whistles. First, friends of Katy showed up, and then eventually Katy. When Stubby arrived I decided to play his whistle and everything changed. Stubby came over, directed an open mouth at the speakers, and swam off. I interpreted this as a message of inappropriateness and felt that we really didn't know what messages we were potentially sending. Were we insulting Stubby? Was it inappropriate to play the signature whistle of a male or of anybody for that matter? At that point I decided that these playback experiments would be unproductive, and perhaps inappropriate to their rules, and I ended our playback work.

Late afternoons we watched dolphins begin their southwest travel to amass at the deepwater edge. Although dolphins can swim fast if they need to, typical spotted dolphin travel is slow and methodical, cruising along at slow speeds of around one to two knots. Their high bursts of speed are associated with herding fish or chasing after an intruding bottlenose dolphin. In the last few years we had observed the spotted dolphins coming in to the shallow sandbank from deep waters in the early morning. Especially after Ken's visit we started to wonder whether the spotted dolphins were more similar to spinner dolphins than we had previously thought. Although the spotted dolphins eat fish during the day on the sandbank, access to the DSL on the edge was looking more and more like the nocturnal offshore hunting behavior of the spinner dolphins in Hawaii.

One night we headed off the edge of the sandbank to try to score some calamari for dinner. We had seen the dolphins regurgitate various fish and squid body parts, including the cartilaginous pens and beaks of squid. But we weren't sure where the dolphins were getting them, since there were reef squid on the bank but they were quite small. As we arrived offshore and drifted in thousand-foot water there were the squid! And there were the spotted dolphins feasting on our visions of calamari. Their large amassed groups at the edge had caught the northward flowing current to forage in the rich DSL that rose to the surface with its booty of slippery squid and flying fish. Foraging on the edge at night is a spotted dolphin activity. The bottlenose dolphins choose the grass beds of the sandbank as their niche, thus dividing up the wealth of the sandbanks between two potentially competing species, leaving the squid to be caught by hungry dolphins and humans alike.

By early August night drifts had become part of our regular activity when weather permitted. We saw every age-class and sex feeding at night, from the tiniest calf to full-grown adults. While adults hunted schools of squid that disappeared into the depths, calves, with their mothers in tow, chased after three-inch flying fish for target practice. More experienced spotted dolphins scored the fatty flying fish. Often mistaken for birds at sea because of their ability to glide in the wind, flying fish became casualties, landing on our boat and injuring themselves. Sometimes they hid against the boat where the dolphins had difficulty grabbing them until the dolphins, determined to dislodge their prey, emitted a large air bubble to scare the fish away from the boat. Flying fish sometimes flipped out of the way and landed behind a hunting dolphin. It is not so easy to catch a flying fish if you are a dolphin because, like most prey, flying fish have specific strategies for escape. Flying fish flew into our heads while we were working in the water; they flew into us while we observed from the deck, always leaving an oily, fishy stench. Calorically, a flying fish is a good fatty meal, and the squid are primarily water. Pantropical spotted dolphins in the Pacific preferentially feed on either flying fish (lactating females) or squid (pregnant females) for reproductive reasons. The first time I ate a flying fish I realized how good they were, a delectable morsel for both humans and dolphins. Not being as talented as the dolphins, we found ourselves limited to the casualties on board for our flying fish meals. Although normally we ended our night drifts by midnight to go back and anchor in the shallows, on occasion we drifted all night passively in the current to see if the dolphins kept feeding until morning. As we passed group after group of spotted dolphins that had set up shop along the edge we realized that the dolphins did drift all night, unlike their human companions.

In mid-August Tropical Storm Bob was forming off Nassau, Bahamas, about three hundred miles southeast from us. Flip got off the boat and headed to New England, glad to be getting out of the way of Bob. We tucked into port as Bob harmlessly went over us and then became a hurricane on its way north. We were lucky once again, but like a shark following a dolphin, Hurricane Bob followed Flip to New England and then went on to Nova Scotia for an East Coast tour.

My birthday is in early September so my birthday celebration usually took place on board our research vessel where it's pretty hard to hide party preparation. Attempts were made to distract me while decorations were put up in the salon or cakes baked. Since beginning my fieldwork, I can't remember having a picnic on land in the summer, or a Fourth of July celebration, or my birthday anywhere but at sea. My birthday month marks not only my increasing age, but also the end of my field season in the Bahamas each year, and the precious time I spend with this wild society. This birthday I watched as Hurricane Claudet formed and headed north of us toward Bermuda and I wondered how many hurricanes I would witness on my birthday in the future, and if they would all miss us out at sea. In the next few years we would not be so lucky.

In the fall of 1991 I gave the first presentation of my work in the Bahamas at the Marine Mammal Conference (MMC) in Chicago. A biennial event, the MMC is the professional conference for researchers in the field of marine mammals. I was anxious to present my work, since I really did have some basic correlations of underwater behavior with vocalizations, and I wanted to share them. In those days we had slides and slide projectors, and luckily the conference allowed two slides at once so I got up on stage and began showing underwater behavior and some video with sound correlations. Later Flip Nicklin walked up to me and said that my talk was "hot." I guess he meant no one had ever been able to document what I did, and it was opening a new window. It was true that by the end of 1991 I did have a sense of some of the patterns of underwater behavior with sound. It was a start and I was glad to share it.

By the end of 1991 I had a good sense of the dolphins' basic behaviors, including some of the "percussive" sounds such as tail slaps and jaw claps, that create loud cracks of sound in the water. Even bubbles make sounds, and dolphins use them for communication signals: whole-bubble expulsions for dislodging fish that seek shelter under our boat during nocturnal feeding, bubble rings as a sign of impending aggression by a male coalition, and bubble trails as a motor marker of a signature whistle or during the loss of vocal control when an excited calf swims erratically and is calmed down by his mother or babysitter. I constantly watched and wondered how many other communication signals I would see in the wild. I would need a few more years.

After the excitement of the conference I found myself more anxious than usual. Our three-year charter with the *Wren of Aln* was up. Doug had decided to move back on board his boat with his family so I was on the hunt for a new boat to charter for the summer. We had been slowly raising money to build a boat and after realizing the real costs of such an endeavor, Dan and I began traveling the state looking for a good multi-hull to purchase. And then a miracle happened.

A long-term friend of the project had mentioned a catamaran he had seen in a waterway near his home. We jumped in a small boat and headed up the waterway. From the water she was beautiful. There she was, the *Bo-Ann,* large, heavy, sixty-two-feet long and twenty-three-feet wide, she was clearly built for open-ocean work. I was anxious to talk to the owners to find out if it might be possible to charter this boat for next summer, so I obtained the address of the home nearby and wrote a letter introducing myself and my work. Two days later I received a phone call from the captain who was living at the house. "Well," he said, "the boat is owned by a corporation and they don't want to charter it because of liability, but they are looking for someone to donate it to." I couldn't believe my ears! Could this really be happening? They wanted to donate it and get a tax write-off? Hmm, the Wild Dolphin Project was a non-profit, 501(c)3 . . . hmm. I quickly called a few board members and told them about the opportunity, but I was equally concerned with the fact that taking on such a boat is a huge budgetary commitment. Owning a boat is like owning a house versus renting, suddenly you are responsible for paying the insurance and fixing the broken air conditioner. We arranged for a sea trial and I remember walking on board the boat, after having viewed many a small sailboat within our price range, but whose space meant that we would not get regular time up in the dolphin area in even moderate weather. I thought, "Wow, I could live on this boat for another ten years!" So, after some fancy negotiating and appraisal, we did what is called a bargain sale donation, we paid the owners some cash and they wrote off the rest of the appraised value.

Acquiring our own catamaran was certainly one of the best things that ever happened to the project, since it secured in many ways our bricks-and-mortar for the project, and to this day still provides the oceangoing platform for our work, but it was one of the more stressful times for us financially. I have had to be a master juggler of money running a nonprofit for so many years. Many people think because you have a long-term project that you have long-term funding. Not true, I just had long-term determination. So every year we struggled for funding. We got grants, we brought the public out with us in limited numbers for ecotourism, and we sold T-shirts. We still do all of these things. When we got the boat, suddenly I had a schedule of payments to make for the bargain sale, and it was stressful. I remember driving home one day, crying, wondering how I would make the next payment since we didn't have the money. Once again, Lisa, Peyton, and other board members jumped in to help with temporary loans, to stay on schedule and to stay afloat.

Built by an aluminum company called Modern Welding in Kentucky, the *Bo-Ann* was a sturdy boat, but she was a bit neglected. Built for the chairman of the board, who then willed her to his children, the *Bo-Ann* had not seen much sea time lately, and the worst thing you can do to a boat is let her sit. So we duct-taped the canvas top on the bridge, and used 2400 (the superglue of boats) and WD-40 to hold the boat together. We sawed the king bed in half to make bunks and put folks on the "couch" in the salon the first year, but gradually through the sweat and blood of a few captains and first mates our vessel, like fine wine, improved over the years.

We renamed her *Stenella* after the genus of the dolphins, and our smaller boat, *frontalis*. My friend Pat, the glass artist, etched us a dolphin windowpane, as the *Bo-Ann*'s stained glass window was removed. And *R/V Stenella* was created. Over the years we got rid of the canvas top and put a hard, molded top on *Stenella,* we built a research table in the salon, improved the bunks, and constructed a large picnic table on the stern for dinner outside.

But right now we had a boat, and that's all we needed. We also had a budget that was double. My work has been blessed in many ways, but beyond everything was the miracle of acquiring *R/V Stenella,* and the series of timely events that took place to get our new boat. As the winter flew by and I geared up for a new era of fieldwork with *R/V Stenella,* I wondered how the dolphins would react. Would they recognize us on a new boat? Would they enjoy bow riding on two hulls instead of one? One thing I knew for sure, this was a boat that was well suited to my field site and the fieldwork, and it was ours.

PART 2

The Middle Years

Observation—1992–96

Cracking the Code: Detection and Deciphering

By this time we had established a relationship with this community of dolphins. They not only tolerated our presence but they were showing us natural behaviors within their own society. It was as close as you could get to seeing natural behavior underwater in the dolphin's world. The dolphins now viewed us with no fear and even trusted us as they went about their business foraging, nursing, and fighting. The investment in the relationship had paid off handsomely.

The mid-1990s were El Niño years, which for us meant a low probability of hurricanes. Stretches of good weather and calm wind made the difference between a great data collecting season and a mediocre one. The dolphins' behavior was complex and piecing together their story was a process. The dolphins showed us intimate details of their lives including how they spent their time during nighttime hours and the complex relationship with the local bottlenose dolphin community. I observed, recorded, and analyzed the dolphins underwater, now recognizing the complex patterns of their behavior. I watched Little Gash and Rosemole raise their first offspring and saw juvenile dolphins practicing for adulthood. I was beginning to "crack the code."

~ 4 ~

Dolphin Society Takes Shape

> It is one of the most striking aspects of chimpanzee society that creatures who can so quickly become roused to frenzies of excitement and aggression can for the most part maintain such relaxed and friendly relations with each other.
>
> —Jane Goodall, In the Shadow of Man

1992—Gangs, Girls, and Games

After acquiring *R/V Stenella* in February, Dan and the crew were trying to get her ready for the summer field season. The priorities had been the engines and propellers; cosmetics fixes would wait. Although I had already been running my project since 1985, and my own nonprofit since 1988, a large boat entailed huge responsibility, you need insurance, you need to fix things, you worry about the structure and, in the case of a boat, potential leaks. But it was our boat and finally I was able to design my own field schedule, hire my own captain, and remodel the boat to function as a real research platform.

May 23 broke clear and crisp for our first day up on the bank and what a reunion! Did they recognize us from the bow on our new boat or were they checking out the new boat with two hulls guaranteed to provide multiple bow rides? Even White Patches, normally mellow and sedate, was excitedly exploring the hull. As we entered the water I greeted Rosemole with her now two-year-old calf, Rosebud, who, like her mother, exuded sweetness. Little Gash was dashing to and fro, flirting with the now interested and courting male dolphins in the group. And much to my relief Gemini was here.

I met Gemini in 1986 while she was still nursing her newborn female calf, Gemer. In addition to other identification marks Gemer had a remora. But in the spring of 1989 Gemer suddenly disappeared. Then one day after anchoring, our cook saw a fused spotted dolphin under our boat chasing needlefish at the dolphin wreck. Captain Dan and a passenger got in the water and quickly noted that it was the sick, extremely thin, and emaciated dolphin we had seen swimming listlessly and alone all day. This dolphin wasn't particularly concerned about people in the water and continued hunting small fish on the wreck. Dan observed her catch a needlefish with great effort. When I got in with the underwater video camera the dolphin came close enough for me to see wounds on her left side and her two remoras. She was disoriented, and approached the camera with her widely spaced echolocation clicks, atypical of hunting sounds. Her blowhole area was also dented, indicating she had been burning her reserve blubber in the front melon area. As she swam around us I had a feeling of discomfort; the dolphin didn't seem aware of us. Later that day the same emaciated dolphin briefly rode the bow. When reviewing videotape later that night I was shocked—it was Gemini! I saw her distinctive ragged dorsal fin and spot patterns, but where was her calf? I could only imagine one scenario. A shark or another predator snagged Gemer. I envisioned Gemini fighting off a shark to save her calf. If so, did Gemer's remora jump ship onto Gemini? On the other hand, Gemini's remoras could be her own property, acquired because of her deteriorating condition. If she wasn't eating and was emaciated, why would a remora stay on her? Was it because she had a new wound with flesh hanging on the edges? Was Gemini's listless behavior a sign of grieving for the loss of her calf?

I observed her later in the season and not only had she survived but she had regained her energy. She now had a telltale white spot under her left pectoral fin, a clear scar from her wound. A decade later Gemini proved a stable member of the community, reproducing and successfully raising more offspring including Geo (1991), Galaxy (1997), and Galileo (2000). Although Galileo was a female I chose her name because many human females in history hid behind male names to get noticed. Witness the scientific life of Madam Curie and others; it seemed appropriate. Galileo was smart, quick, and spirited. Her older sister, Galaxy, was a sweetheart and her brother Geo was a stable presence.

A striking example of remora behavior occurred with one of my favorite dolphins, Heaven, who was presented as a newborn to me by Hedley, her mother, one late afternoon in 1995. After a physically exhausting day with the dolphins, and many hours in the water, we anchored our research vessel. A group of dolphins suddenly passed by our boat. Hedley and her new calf, and another mother with a small calf in tow, swam by. Lynn Thomas, the cook, and I slipped in the water, unable to resist another look at such a small calf. Hedley slowly came over to show off her new offspring who still had the telltale visible fetal folds of a newborn. The young calf made a few whistles, and she was bold for her age, possibly two to three weeks old, and already sporting a great identification mark behind her dorsal fin. I felt grateful that Hedley had brought her newborn calf over to our anchored boat. She presented her calf to us as if with great pride. I named her calf Heaven, inspired by those strikingly beautiful wisps of white and gray on her sides. She was truly heavenly.

Four years later, in 1999, we saw Heaven, by then a juvenile, with a serious skin problem. Heaven was a very sociable juvenile dolphin but I only saw her alone this summer, and she was often whistling sadly as if searching for companions. This particular August day I remember thinking how unusual it was to see a single spotted dolphin alone. When I entered the water it was clearly Heaven, slightly listless in the water. Her skin was extremely discolored. On her right side she had cysts and muscle ridges were visible where fat and skin had thinned. She was making her signature whistles, and dragging a piece of sargassum, but she wouldn't play. The change in her skin had dramatically worsened from the last time I saw her just four days before. She seemed to hang around the boat as we drifted. Sick dolphins are often seen alone; perhaps by choice, perhaps they are ostracized. Jane Goodall, in her long-term chimpanzee study, observed that sick chimpanzees are often shunned from the group, sometimes even killed. Heaven's whistles made me think that she was trying to make contact with her pod but no one was responding. Or perhaps she was just lost or losing her senses and becoming disoriented? By the end of that field season I tried to accept the fact that Heaven would probably not survive.

Next spring, much to my surprise, we saw Heaven and her skin looked beautiful, clean, clear, and healthy. She still had some skin patches and a dark wound on her right side, but not anything like last year's cysts. And there was no remora. I often empathized with the dolphins and how they must feel about their annoying remoras. In fact, my research team plotted and brainstormed about trying to remove remoras from the dolphins by grabbing on to them or luring them away. But in hindsight, I now realize that for the most part these remoras might be helping the dolphins by removing dead skin as they travel up and down the dolphin's body. It's a good reminder to not interfere with nature when we don't understand the complexity of relationships.

1992 heralded in a stretch of great behavioral observation years. In addition to having our own boat, and having good weather, I now knew many of the dolphins and I could readily identify their developmental stages by their spot patterns. The more comfortable the dolphins were with us the more they displayed their normal behavior. It is impossible to be totally invisible to any species. On land you can build a blind to try and hide your presence. In the water all we could do is act benign and respectful so the dolphins went about their normal business. These years were rich with intimate and regular observations underwater.

The shift in the dolphins' relationship with us was incredible and I couldn't imagine what the next five years would be like. What does it mean to have two such disparate species connect? What's in it for the dolphins? Being incorporated into a society of wild dolphins was akin to an anthropologist interacting with a new human culture and learning the signals and habits of that society. I had the overpowering feeling that the dolphins now expected us to recognize and respect their signals and patterns and act appropriately in the water.

One example where the dolphins showed their recognition of the human situation was in the discipline of their young. Calves, excitable as they are, sometimes get out of control even when tended by a responsible babysitter. Although young calves may not understand safety boundaries at this age it is also possible that they are testing their boundaries with the adults. During one swim, Katy, a young female babysitter, came by me and swished and slapped her tail in front of my face as if to say, "You're getting these guys too excited and it's my job to control the situation." I quickly stopped paying attention to the calves. Later a mottled male, Big Wave, chased a postpartum female to mate with her while people unintentionally interrupted his attempts by getting too close. Big Wave swam by me swiping his tail near my face to communicate, "Yield to me my space to perform my social function." Without trying to be anthropomorphic, that is the message I got. Although it could be interpreted as aggression by an inexperienced eye, to me it was clearly the transmission of their social rules and an assessment of their, and our, behavior. I believe this is how they communicate rules in their own society. It is also interesting that the dolphins often address these signals to me, possibly because I am always in the water and it may appear that I supervise these other humans to a certain extent (which I do). That would mean they recognize roles and responsibilities in another species and direct their communication appropriately. My colleague Dr. Thomas White and I wrote a paper later about these scenarios and how they add to the list of qualities of "personhood" for nonhuman species. It is a sophisticated skill to say the least.

I have never regretted taking time to learn their rules at the expense of data collection. It was the best investment I made in the research to ensure its continuity. We were constantly tested on dolphin social etiquette—if we passed, more secrets were revealed, if not, it was back to square one. Engaging the dolphins as full participants in the work, and in a personal relationship, has been my primary approach over the years because without the dolphins' cooperation, as partners in this learning, we were nowhere. Our use of benign methods, instead of tagging or taking skin plug biopsies for genetic work, preserved the unique opportunity at this field site, which is first and foremost to observe the dolphins underwater.

For the first five years it was probably me who adjusted to the dolphins more than the dolphins to humans. The dolphins remained themselves and did their normal activities. I, on the other hand, changed my behavior in the water, my approach angle, and the way I would vocalize or use my hands in the water, all to gain their trust. So the reverse is probably true: in our interactions with animals we adjust our own behavior in order to observe the other. Of course if we truly believe it is interaction, not pure observation, then we must admit that we, too, changed the behavior of the dolphins. In addition to tolerating us while they went about their business, the dolphins learned our behavior to their advantage. The dolphins stationed themselves at our bow while we worked with a distant group, knowing that when the boat eventually moved over to the humans in the water they would gain a bow ride. The dolphins often tail-slapped at the bow to get our attention and to signal their desire to bow ride and we willingly complied. When we enter into the world of other beings we become susceptible to their rules and subject to both the responsibilities and rewards of such a relationship.

Over the summer we moved the boat around more during the day than in previous years since we felt that the dolphins were, by now, fairly comfortable with us. We hoped to see other behaviors than the ones we observed while anchored. And I was right. When we were at anchor we got glimpses of their behavior, brief snippets, as they traveled along. As a now mobile vessel we drifted with them for hours; we joined in their travels and observed a normal dolphin day.

Babysitting is one of the most striking and important aspects of growing up dolphin. Katy, a developing young female, had regular babysitting responsibilities. This day, after giving some juveniles a bow ride, we entered the water to observe a young calf with his rostrum near to Katy's genital area. The beak-genital orientation is something we usually see during courtship behavior when males are inspecting and potentially stimulating a female. This behavior can appear like nursing and I was reminded of Apollo, the first calf of Luna, who I saw beak-genital with Little Gash when she was babysitting this youngster. As I watched the scene unfold I wondered if young calves mistake a babysitter for their mother or whether the calf is just seeking comfort or pleasure.

A few days later we traveled south and picked up Blotches and her calf Baby with Stubby, Little Gash, and Luna. Suddenly I saw and heard White Patches (I could recognize her distinctive signature whistles at a distance) rapidly round up the calves. Like Katy, White Patches was of babysitting age and, more than any other dolphin, she was the quintessential aunt. She darted in and out of the group of young dolphins, rounding them up one by one as I recorded multiple, and overlapping, signature whistles, which seemed to belong to the calves. We know from captive research that dolphins sometimes use another dolphin's signature whistle to make contact; and they sometimes expel bubbles when vocalizing, but in the wild it is difficult to sort out. But when I saw bubbles coming out of her blowhole at the exact time when a whistle was produced, it appeared clear that White Patches was using the calves' names (signature whistles) to gather them together, in addition to her persuasive swimming and herding.

Little Gash, as a now young mottled female, engaged in different activities this summer. This June day Little Gash appeared with a group of mothers and calves, including Gemini and Geo, scanning and foraging on the bottom. I thought it strange that Little Gash was hanging around foraging moms instead of her normal female friends, but in fact Little Gash was newly pregnant. Was Little Gash in need of extra food and exposure to mothers with their calves? We were starting to see how dramatically females changed their associations based on their reproductive cycle. In some cases pregnant females from the southern group associated with pregnant females from a completely different area, and that was what Little Gash was doing today.

The males, too, had developmental indicators and patterns that began to emerge, but these were opposite the female patterns. Although most of the juveniles I knew by now were female there were a few juvenile males in the group. There was Poindexter with the bumpy leading edge of his dorsal fin, Havana with a large shark bite out of his fluke, Horseshoe with his mark on the left tail stock, and Geo, who always seemed to be low on the totem pole and bore a nasty remora and had skin problems for most of his life. These young males practiced the sequences of mating behavior, although awkwardly, as a young female, Uno, found out this day. Poindexter was mating with Uno when another young male, Warbler, approached and began a head-to-head posture, usually an aggressive indicator, with Poindexter. Juvenile aggression takes on a casual pace unlike escalated adult aggression, which is fast and violent. These two young males interwove their movements in a graceful figure-eight pattern, with occasional squawks and charges. Then Poindexter and Warbler both began stimulating Uno with a beak-genital buzz. Even at this early age the building blocks of male coalition cooperation were evident. These juvenile male spotted dolphins spent their time differently than females; they fought, played, and practiced courtship rituals with each other and any willing juvenile female they found. It is during these juvenile teenage years where their lifelong relationships are forged and solidified.

Another striking aspect of growing up male was the mock fighting sessions between older males and younger males. I had seen a glimpse of this before in 1991 when I observed Stubby one day helping some juveniles hone their foraging skills. But this August day Stubby was the only adult male present while Havana and Poindexter frolicked with a few mother-calf pairs. Blotches, Baby's mom, and Havana's mom, Hedley, with her deteriorating right eye, were there. Stubby went head-to-head and open mouth with Blotches and Hedley as the two young males, Havana and Poindexter, went after Baby, attempting to copulate with her. Havana approached Stubby and postured head-to-head aggressively, as Stubby gently encouraged the subtle adjustments of his signals during this aggressive standoff. Displeased and ready to discipline Havana, Hedley chased Havana in the upside-down inverted posture mothers often do, but favoring her good eye Hedley looked more like a pirate than a frustrated mother.

By the end of the season I had yet another example of a male teaching Havana how to mate. Shorty, a very old fused male and recognizable by his indented dorsal fin, approached Havana with an *S* posture, a genital buzz, and then copulated with him. Shorty swam over, while Havana watched, and postured, buzzed, and copulated with Havana's mother, Hedley. Well, there was nothing like a clear demonstration. Then Havana went through the same drill and copulated with his mother. Of course a human parent might be aghast at such teaching methods, but for dolphins sex is very much a part of their social interaction and it functions not only for procreative purposes but also for social bonding. Mothers with newborn male calves prod and encourage erections and copulation at a young age. When mothers with two-year-old calves are courted and approached by male coalitions the young calf is attending, watching, and hearing all that occurs. It is striking how much exposure a young dolphin gets to social signals the first few months of life. So it is not surprising that stranded young dolphins, or dolphins born in captivity, do not have the exposure or social skills to reproduce or interact normally. In a wild dolphin it takes many years to learn and practice appropriate social skills.

Dolphins also display signs of jealousy during their interactions. By the end of 1992 we had spent time in the northern area of our study site, which meant time to get to know the smaller northern group with only a few family lines. Snowflake and her male offspring Sleet (who had a cataract in one eye), and female offspring Snow, were there in addition to Slice, a young male. While we were interacting with Snow, Trimy, a young juvenile female, started tail-slapping when Snow brought over a piece of seaweed to solicit play with us. When Snow backed off from her human solicitation of attention, Trimy continued her interaction with us, and it occurred to me that I was seeing my first example of a jealous dolphin. And it would not be the last.

Pouting and peacekeeping are also a part of a dolphin's social repertoire. It was mid-September, the end of our field season, and a small group of juvenile dolphins were roughhousing and chasing each other with head-to-head displays, the roots of aggressive behavior. Zigzag, a young male, apparently distraught over the escalating activities and after evasive and exhausting maneuvering to escape, made little chirps and hung on the surface making in-air raspberries (little whimpers from the blowhole heard in the air). Only after the other juveniles gently approached and rubbed him did Zigzag rejoin the activities. Now think about this; a dolphin is distraught because his buddies are too aggressive and he is afraid and pouting. His buddies recognize his distress and know exactly the apologetic gestures to invite him back in the game! This is an incredible example of the dolphins' advanced social abilities and empathy. It would be impossible to not be self-aware and empathic for such an event to occur. There is no doubt in my mind that dolphins are both.

The extent of the development and challenges of a dolphin's social life became apparent this summer. In the middle of September, Paint's young calf Brush showed up with a large, and quite fresh, shark bite. To this day her scar is visible even though she has been a mother multiple times and her mother, Grandma Paint, still has offspring. On this late summer day Paint seemed protective of her first calf and startled when I approached too close in the water. As part of the northern group, Paint and Brush probably had less protection than other females, such as those in the central group. Networking can make a difference and it has been observed that chimpanzee mothers of lower rank who live on the "periphery" of the best foraging grounds are often more challenged and less protected than females of different lineage or higher rank. I suspect it is so in the dolphin world, although it remains unclear how they both acquire and display such rank. No wonder dolphins have strong social bonds; their lives depend on it.

By the end of the summer I had seen how important it was to watch behavior develop over time within this society. Understanding the pro-

Rosemole, one of the first dolphins I met in 1985, swims along with her first offspring, Rosebud. Like many first-time mothers, Rosemole was successful probably thanks to her experience as a babysitter in the community. *(Denise Herzing, Wild Dolphin Project)*

Well-known for his chopped-off dorsal fin, Stubby has been one of the most regular dolphins in the community. *(Michelle Green, Wild Dolphin Project)*

Female spotted dolphins form strong friendships when they are juveniles, but this changes when they begin reproducing. Once Rosemole and Little Gash became synchronized in their birth years, their off-spring Rosebud and Little Hali became fast friends. *(Will Engleby, Wild Dolphin Project)*

Little Gash and her calf Laguna swim along during a nursing bout. Dolphin milk is rich and provides good nutrition for a growing calf. *(Will Engleby, Wild Dolphin Project)*

Dolphins display aggression using head-to-head posturing and open-mouth displays. *(Denise Herzing, Wild Dolphin Project)*

A juvenile dolphin hovers vertically in the water column while opening his mouth during play aggression. *(Denise Herzing, Wild Dolphin Project)*

Latitude tries to take cover with an older group of males while his mother chases him inverted during her attempt to discipline him. *(Nicole Matlack, Wild Dolphin Project)*

A group of Atlantic spotted dolphins hold down another on the bottom. Mothers hold their calves down on the bottom in the same way for discipline. *(Will Engleby, Wild Dolphin Project)*

During the courtship ritual male dolphins bend their bodies into an S shape before approaching the female inverted and from underneath. *(Denise Herzing, Wild Dolphin Project)*

Dolphins are master acoustic and postural mimics and use this ability to communicate. *(Will Engleby, Wild Dolphin Project)*

Touch is frequently used for communication between dolphins. Pec-to-pec, pec-to-head, and pec-to-body rubbing are used during play and reconciliation behavior. *(Denise Herzing, Wild Dolphin Project)*

In the wild dolphins create bubble rings that signal impending aggression. *(Will Engleby, Wild Dolphin Project)*

Little Gash and Tink have target practice with an unwilling fish. *(Marsha Coates, Wild Dolphin Project)*

Atlantic spotted dolphins in the Bahamas regularly dig in the bottom for flounder, razorfish, and eels. *(Denise Herzing, Wild Dolphin Project)*

Although typically individual hunters, Atlantic spotted dolphins occasionally cooperate and herd fish balls as a food source. *(Denise Herzing, Wild Dolphin Project)*

Remora australis, a bluish-purple remora, is found attached to only whales and dolphins. With suction created from its modified dorsal fin, remoras hang on to their host. *(Denise Herzing, Wild Dolphin Project)*

Sharks attack both adult and young dolphins. Nose, a female bottlenose dolphin, survived this attack but still bears the scars. *(Denise Herzing, Wild Dolphin Project)*

Two male bottlenose dolphins side-mount a male juvenile spotted dolphin. *(Will Engleby, Wild Dolphin Project)*

Atlantic spotted dolphins mate belly to belly, with the male underneath the female. Male coalitions herd and monopolize a female in estrus and take turns mating with her. Juveniles also practice mating as part of the learning process for their adulthood. *(Will Engleby, Wild Dolphin Project)*

Katy, daughter of Blotches, and mother of our first third-generation dolphin, Kai, swims along in a visibly pregnant state, followed by the author and her video camera. *(Will Engleby, Wild Dolphin Project)*

Research Vessel *Stenella*, a 62-foot power catamaran, has been the research platform since 1992. Strong and seaworthy, *Stenella* serves as offshore laboratory and home during the summer field season. *(Will Engleby, Wild Dolphin Project)*

An underwater keyboard and jar full of dolphin toys are used in an attempt at two-way communication. *(Marsha Coates, Wild Dolphin Project)*

During interactive encounters and natural behavior, video recordings with sound are made for later analysis. *(Michelle Green, Wild Dolphin Project)*

Little Gash mimics a diver by lying on the bottom. Spontaneous and creative behavior is not uncommon during human/dolphin interactions. *(Denise Herzing, Wild Dolphin Project)*

Caroh, our main ambassador for interactive work, was notorious for stealing scarf toys from humans. *(Courtesy of Ruth Petzold)*

cess of the development of communication signals told us much about a normal healthy dolphin. Two months ago Little Gash was barely showing, but now she was clearly pregnant with her first calf, although not quite as large as Nippy looked last fall before we left. Now she could catch up with her good friend Rosemole who already had a calf. Would their calves become friends or would the reproductively driven separation between Rosemole and Little Gash affect the potential calf relations? As we watched through the years ahead we saw the mothers' relationship wax and wane as they went in and out of different reproductive stages. Never was the dynamic of this relationship so pronounced as when their third friend, Mugsy, lost her calf.

By the end of the 1990, both Rosemole and Mugsy were pregnant for the first time. When I left in the fall I remember thinking how exciting it would be to come back next spring and see both these females with their first calves. When we returned to our field site in May there in the group were both Rosemole and Mugsy. Although Rosemole had a healthy young female baby, Rosebud, and was surrounded by other mothers proudly swimming along with their calves, Mugsy, in formation, was despondent and without a calf. Perhaps it was a miscarriage or she might have lost her calf for either natural reasons or to a shark. Like primates, dolphins have been known to carry around their dead offspring for days. Mugsy swam in line with the other mothers, leaving a space underneath her as if she had a phantom calf in tow. Month after month she showed no interest in her other friends and new mating opportunities. The trail of grief for her lost offspring was palpable.

By the end of my field season I realized that I was not only lucky to have found a place on the planet to watch dolphins in clear water but also to be working with a species whose age classes and developmental phases are so clear and pronounced. Observing the juvenile equivalent of adult behavior was an important aspect of understanding and interpreting their underwater behavior. They made mistakes, had their signals adjusted or encouraged by adults, and practiced. But how did they learn things? Was it by trial and error, imitation, or instruction? I had seen bits of all these things and I suspected that there were many ways they learned. Learning also depends on their tutors and mentors, including the bottlenose dolphins that are part of their world.

In early August I had observed an interesting example of juvenile sexual play that appeared to be an imitation of interspecies aggression. Diamond, Luna's female calf, was the focus of most of the activity. Two juvenile males approached her, using their normal genital buzzes and orientations, but then, as bottlenose dolphins do with their females, both spotted males started mounting Diamond from opposite sides at the same time. The males used head-to-heads, open mouths, and tail slaps while they squawked at the uncooperative Diamond. I wondered if this was juvenile courtship practice gone bad or if the bottlenose dolphin behavior was something that young spotted dolphins were not only observing but had learned to mimic. I could tell that the interaction between these two species was going to be critical to the understanding of the lives of these wild dolphins.

It's easy to see how dolphins socialize, rest, and play in the safe shallows of the sandbank during the day, but it is another thing to see them in eight hundred feet off the edge of the sandbank, in the dark night, hunting for food. It is here where we see what the dolphins' night life is like, and where we, as human researchers, are most inept. If you want to feel what it is like to be a dolphin, get in the dark water at night and swim away from the boat. When depth is only determined by a depth finder, or a dolphin's sonar, you will soon discover how scary a place it can be. This is why dolphins that live in deep water are group animals. The game changes when the depths below look infinite and you are unable to distinguish a large predator on the bottom as you are on the white sandbanks during the day.

Because of good weather this summer we often drifted offshore at night, which takes the right set of circumstances. The weather has to be relatively calm, the waters flat, and the tidal cycle not too extreme, since it will move the boat either back on to the sandbank or further offshore. Often we cruised down the edge of the bank at sunset and if we were lucky we ran into groups of dolphins amassing at the edge. They played and dawdled, if a dolphin can dawdle, until the sun went down. Then, in a larger group of twenty or so they moved off the shallow edge. The northward flowing current along the edge of the sandbank allowed us, and probably the dolphins, to use minimal energy to drift north. It is not exactly clear why we would find them in certain locations off the edge. The sandbank is actually a diverse habitat with certain areas sporting beautiful reefs near the edge. Here the DSL rises, bringing a diversity of organisms to the surface. These areas were also our best fishing grounds when we trolled offshore for dinner. I imagine the dolphins have much greater knowledge of these subtleties and of course use them to their advantage. About half the time we went offshore we found the dolphins, the other half we did not. We ourselves were limited by the weather, our samples opportunistic. When we found the dolphins on the edge we watched them for hours. We turned off our engines and drifted with our hydrophone overboard listening for their telltale clicks and whistles. Their prey, schools of six-inch squid or eight-inch flying fish, appeared first. On occasion needlefish or ballyhoo came by and we dipped a net in the water to snag a few and look at their rainbow of colors in a small bucket. Thirty seconds after we heard the first echolocation clicks, the dolphins appeared. Click—click—click—cl-cl-c-c-c-c: the rapid rates of clicks increased as the dolphins flew underwater after their prey. Although our deck lights probably helped attract their prey we sometimes left our lights off for a drift and the dolphins still appeared, audible through our hydrophone underwater, audible with their breaths at the surface, and visible with our nightscope and available moonlight. Sometimes we turned our deck lights out to see if the dolphins would leave, but they didn't, they continued to feast for hours. As we started the engines to leave and go anchor in the shallows a few dolphins gathered at the bow, leaving their midnight buffet behind. We watched their glistening streamlined bodies gliding on the bow, outlined by the starlike bioluminescent organisms that flow and twirl around their torpedo shapes. It was elegant and breathtaking, like the twinkle of stars in the sky; the flashes of light set the dolphins off in a universe of their own in their watery galaxy of the deep.

One night we again ran south and started back up the edge to look for foraging dolphins. We found Havana with Katy socializing near the edge. Flash and Scqew, two up-and-coming males, were playing nearby. We were in three hundred feet trying to stay with this group as they headed off the edge for their nighttime hunt. We traveled up the edge in a hundred feet and saw some subtle movements at the surface offshore. There were Uno, Hedley, Havana, and Little Gash already in 477 feet. White Patches and Katy showed up and seemed excited, making signature whistles and excitement vocalizations as they rallied their troops offshore. I thought of Ken Norris and the parallels of his spinner dolphins heading offshore to feed in Hawaiian waters. As they frolicked with seaweed in this deep water the spotted dolphins sent their clicks of sonar toward the bottom, visualizing creatures that we could only imagine. Tonight our depth finder showed us a great reef wall five hundred to a thousand feet down. We continued to drift and watch the dolphins hunt while we dreamed of what treasures of fish and squid swam below beyond our senses.

Our fieldwork in the summer was punctuated because of weather. Sometimes we had rain and high winds for days or weeks and we took shelter in a protected area to pass the time. When the weather was good we rose at 6:00 A.M. to begin our research. Sometimes we were with dolphins all day, and then drifted offshore at night, only to rise again the next day and do it all over again. As productive as this was, it was also very physically exhausting.

As mid-August approached we watched Hurricane Andrew strengthen and take a course due west toward us. This storm was a monster, large and a category 5. As we ran back to Florida to drop off our crew and haul the boat out of the water I wondered if our boat would survive. Could we so soon lose our new research vessel to a hurricane? These thoughts raced in my mind as I huddled in the dark with my cats listening to the angry storm outside. At the last minute Andrew veered south and directly hit the Miami area well to the south of us and we were spared. So another early September birthday was spent at sea, grateful that we were still afloat and had our first field season with our new research vessel.

1993—Growing Up Dolphin

The first year with our new boat had gone well. She hadn't sunk, the engines kept running, and we were relatively comfortable. The galley was above deck, making it much more pleasant for our many volunteer cooks. Feeding a boat of twelve hungry people, three times a day, is no easy task. In fact it is the second hardest job on board the boat other than the captain's. But the opportunity to volunteer as cook in exchange for spending some time with the dolphins was an easy trade and over the years we had many dedicated volunteer cooks.

OSE still had a vessel out in the study area, but different students and researchers came and went after they realized how difficult this fieldwork really was, despite the lovely images generated from our most spectacular days. Now that my work was public and becoming known, I soon discovered that there was often someone who tried to take credit for the work. Some attached their name to my Bahamian permit without my knowledge, another used the name Wild Dolphin Project in projects as far away as Hawaii, and others tried to scientifically ignore my work, even though aware of it, by not citing it in their own scientific papers when appropriate. It is a harsh reality that I still see in my field and professional life. This behavior does a disservice to science and sets a bad example for students.

I was just finishing my Ph.D. in 1993 and although I had presented my work at a professional conference, I hadn't published in any peer-reviewed journals because I thought it too premature. OSE still advertised their Bahamas "dolphin research" in a way that suggested they had some continuity with the work since 1985 (which they didn't), causing me to have a lawyer contact them with a polite demand to reword their advertisements to be more accurate. They enlisted a university to start a program and brought on graduate student after graduate student who came and went. Long-term tracking and behavioral analysis is difficult on an ecotourism platform with limited field time; it is simply not conducive toward long-term work. Over the years I had even tried to give suggestions to their students or colleagues about a reasonable project to try to accomplish in two or three years, but my suggestions went unheard. I knew, from firsthand experience and eight years in the field, what it took to sex, identify, and behaviorally document this group of dolphins. Without a degree or publications, being unacknowledged was potentially a problem, although it turns out this was my life's work and I am quite stubborn. In fact, perseverance in the face of adversity, and I have had a bit, is one of my greatest qualities, so I wasn't going anywhere. However, the lack of professional ethics and the incidents that resulted were a hard lesson in the politics of science.

But today it was May 8 and the start of another field season, time to concentrate on fieldwork, not politics. Quickly we found our group of regular dolphins including Stubby, Hedley, and Gemini. It was always a relief to see that the dolphins had survived over the winter months. The next few days we saw Hedley with Havana and Nassau and some other two-tones. As suspected, Hedley's right eye was completely gone, and she clearly favored the left side as she swam by us. We watched as Pair One, Poindexter's mom, taught her new two-month-old calf to chase flounder. He was small and skinny, with the telltale profile of a newborn, unlike the filled-out girth of a two-year-old. Within the first year the length of a calf reaches that of a two-year-old, which is about three-quarters of the length of the mother's body, but the calf is not filled out in girth. The next day we saw the northern group with Snowflake and her offspring Snow and Sleet with Pair One and her newborn, Providence. Sleet still had his cataract and was still nursing. Providence nursed and I watched as his mouth explored around the mammary slits of his mother with his anxious and protruded tongue. This I captured on video, a rare sight in the wild, but with our close-up access we had such an occasional privilege. It was going to be a good field season.

In late May we finally got a good look at Geo's dorsal fin with its rows of teeth marks and jagged edges down the backside; the telltale mark of a shark bite. Much to our relief Geo had survived. It is the dolphins that we don't see that are likely victims of more severe bites.

On the northern end of our study area we found Paint, Brush, and Trimy, a young female with three spots under her lower jaw and a small nick in her left pectoral fin. These marks are subtle and not easily detected by the first-time observer. But Sara Earhart, the thirteen-year-old daughter of Anne Earhart, a new member of the project, exited the water after seeing Trimy for the second time and clearly described the marks, an incredible observation. With a clear propensity for science, Sara went on to study the marine sciences and policy making. Her mother, a strong model of caring about the planet and advocacy, was a prime example of modeling the right action in the human world, and it had been passed on to Sara and then later, by example, to Anne's son Nico. Modeling seems effective in both the human and the dolphin world.

One early June night we suspected we might find the dolphins feeding offshore so we drifted with our lights to a depth of 1,200 feet. In eight hundred feet Stubby and Horseshoe showed up, easily identified from the deck. Eventually more males showed up and then White Patches and Rosemole (who was pregnant and with her two-year-old calf). It looked like the younger females took turns babysitting while all the mothers fed. Luna showed up with her two-year-old, Latitude, and the males were catching fish with him, another example of teaching by males. It was interesting that Rosemole and Luna, two very pregnant dolphins, were out here in the deep water. The waters teem with bar jacks, needlefish, ballyhoo, squid, and flying fish; it seems quite the feast, so who wouldn't be here?

In late June we met up with a group of dolphins off the Dry Bar, the shallow reef behind which we take shelter in bad weather. White Spot energetically herded a female with his friend Horseshoe and as we followed south another male joined. As we looked closely we saw a tattoolike mark on his skin. At first I thought it was a fungus or cookie cutter shark bite but this dolphin had poxvirus, also known as the tattoo virus. By the end of the summer we were seeing all sorts of skin diseases on the dolphins. And 1993 was the start of a few very busy boat years up on the bank, which likely acted as an additional stress factor for the dolphins. Like other mammals, dolphins carry bacteria, fungi, and viruses, and stressors in the environment, as in humans, can set off the full-blown disease. This would not be the last skin issue of the summer. In late June Duet, a young male, had a strange pigmentation pattern on his skin, which I described as a salt-and-pepper look. I had seen it on some of the new calves this year but on Duet it was beyond obvious. I would find out later, after showing these images to the well-known marine mammal vet Joe Geraci, that these were from a likely nonerupted herpes virus. What was happening? Was there a source of pollution suppressing the immune systems of the dolphins or did they always have these viruses and bacteria in their system, which were now set off with stresses by dive boat traffic and noise pollution? If so, it was going to be a long summer.

It was so busy this summer with boat traffic that even the dolphins developed strategies to hide from boats. With our half-mile distance rule for boats we tried to not disturb one another. One particular day our boat and the *Bottomtime II* were anchored within the agreed-upon half-mile distance. I watched in horror as a group of twenty dolphins took shelter and literally hid between our two anchored boats as if they knew it was a safe zone from the other boat chasing them. No boat would go in between our two anchored vessels since we were about a half mile apart. I watched the dolphins wait patiently between our two boats as the other boat cruised by, probably wondering where this large group of spotted dolphins had gone. It might have seemed as if the dolphins opened a giant portal and disappeared. I watched the chasing boat give up and leave the area, and the dolphins drifted slowly in the direction from which they came, having found temporary shelter between our two anchored boats. No wonder the dolphins were getting sick. Dolphins can also catch diseases from people, another reason that we tried not to encourage touching these wild animals. But the oceans themselves are, these days, quite sick. Turtles, dolphins, and other sea creatures have the papilloma virus, herpes, lobomycosis (a fish fungus), and other things unnamed. And yes, dolphins can be loved to death. Although not intentionally disruptive, our mere presence, including boat noise and people wanting to observe whales and dolphins, can impact them in unhealthy ways. In the Pacific Northwest in the United States, where thousands of people flock to see orcas, wild killer whales, on a given day hundreds of boats from both Canada and the United States can be seen following a small pod of whales. The experience of observing wildlife is attractive to people, well-meaning people, yet the impact can sometimes be devastating to the animals.

Nevertheless, dolphin life went on, skin problems or not, and we watched with excitement more examples of normal developmental behavior with the dolphins. White Patches, in her normal aunt behavior, diligently tended Diamond and other calves. Katy, as the "elder" seven-year-old, was in charge and practicing her babysitting skills with gusto. A large group of mothers chased snakefish and, while their calves scored an occasional flounder, White Patches blew a large half bubble (a bubble with a flat bottom but round top) underwater, apparently exasperated trying to keep the rambunctious calves in line. Babysitting was not an easy job even for a dolphin as experienced as White Patches. I watched as one dolphin involuntarily convulsed in the water, regurgitating part of his recent meal. As the excitement continued I scooped up the regurgitated squid beaks and pens, adding to my collection of precious samples of vertebrae and fish parts I had collected over the years. Nearby I watched a young male, Dash, as he chased and tried to round up a group of young, vocal males, and I wondered if we were biased in our definition of "babysitting." Essentially Dash was doing what Katy was, rounding up young dolphins that were out of control.

A few weeks later we saw Dash again as he tried to get young Geo's frantic swimming and vocalizing under control. An older group of mottled males were attempting to hold down Geo and Latitude on the bottom, usually a disciplinary action. I wondered if older males were specifically in charge of discipline in young males? Once again we observed Dash engaging with the male juveniles as they attempted synchronizing their behavior and emitted bubbles. I realized that I was seeing the roots of synchronized male coalition behavior. Old adults used the exquisite tactic of eloquently synchronizing their physical postures and vocalizations. But here was an example of young males who practiced but struggled with their synchrony, stumbling awkwardly through their paces. In some ways it didn't surprise me since I had been seeing that the vast amount of exposure young dolphins got before becoming adults was by immersion and practicing behavior, sometimes with supervision and sometimes on their own.

As the summer progressed Katy continued her babysitting responsibilities with newborns of the year. Bishu (Blotches's newborn) and Lucaya (Dos's newborn) were fairly independent from their mothers considering their age, but Katy monitored them as their mothers foraged nearby. Dos occasionally nudged Lucaya's genital area, causing him to get an erection, and then allowed him to copulate with her. Sexual behavior is learned at a young age and functions as a social behavior when sexually immature dolphins are too young to mate. It appeared that such early sexual stimulation was not only normal but was actively encouraged by the mother or the babysitter. Uno, another young female, taught Providence and Baby how to catch fish. Sometimes Katy assisted Uno in demonstrating to the calves the fine art of hunting fish. It was fascinating. Not only did mothers and other adults instruct calves, but teenagers also taught younger dolphins, sometimes siblings, sometimes not. It was clear that the calves had multiple opportunities to learn in this dolphin community.

How does information transfer in a dolphin society? Is it only from mother to calf or does it cut across age boundaries? It turns out that the social transmission of information can be vertical (mother to offspring), horizontal (juvenile to juvenile), or oblique (juvenile to calf, adult to juvenile). All methods have their pros and cons. Vertical transmission, by its nature, is conservative and proven, one of the pros. An experienced mother teaches her calf how to catch fish. The negative is that in a rapidly changing environment, say an extreme habitat shift of prey, traditional techniques might not be adequate. Horizontal transmission, where juveniles teach each other, is nonproven and can be maladaptive. Witness in our own human culture the bad juvenile habits of crime learned in gangs. In the positive, it helps information transmit rapidly through a culture if needed, and the same can be said for oblique transmission where uncles teach their nephews and so on. Any behavior can be maladaptive depending on the individual and circumstances. In our human culture the sharing of bad habits can become firmly ingrained in our societies with no obvious positive adaptation. In the dolphin culture there are many avenues to teaching and exposure. Perhaps some are adaptive, some not, and others are yet to be determined. This summer we saw example after example of the teaching by males, babysitting instructions, and alternative role-playing. By this time I already had a well-developed sense of normal adult behavior, but watching the process of behavior develop in real time was priceless.

This year Little Gash's reproductive cycle finally caught up with Rosemole's and they became fast friends again, and so were their calves. Today Little Hali and Rosebud were getting foraging lessons. As Little Gash chased up a fish for Little Hali, who then spent minutes chasing the tiny razorfish, Rosebud proudly caught a large snakefish after an intensive chase. After all Rosebud had a few years on Little Hali and it showed in her fish-catching skills. Rosemole, pregnant again, stretched lazily in a flexion posture (similar to an *S* posture but with the tail up instead of down). Then Rosemole swam by with a large jack in her mouth, the size and type of fish (in addition to large mackerel) that we see only very pregnant females eat, probably providing much needed calories during gestation. Perhaps it's the dolphin's version of Ben and Jerry's ice cream.

Clearly females associated by reproductive status and by the end of the summer I was beginning to see how the dynamics worked. Even though Rosemole had, in her first pregnancy, changed her allegiances because of caloric needs (and possibly exposure to motherhood), when Little Gash and Mugsy became mothers, Rosemole was happily back with her good friends. But by late summer Rosemole was once again very pregnant and was now associating with pregnant females Pair Two and Stoplight, two northern females that Rosemole barely knew. Although the friendships among Rosemole, Little Gash, and Mugsy were occasionally renewed when their cycles were similar, their friendships were never as strong as during their teenage years.

My early observations of engaged and competent mothers like Luna, and her friend Gemini, led me to believe that most female dolphins were destined to be successful mothers. Many females, like Rosemole, Little Gash, and Mugsy, spent much of their juvenile years as babysitters and with this training went on to be successful new mothers. But I was to learn that this is not always the case: there are female dolphins that do not give birth but provide excellent caretaking, and there are mothers who are not capable caretakers.

Females that don't reproduce (for whatever reason) often become the society's babysitters. Dolphins like White Patches come to mind. I had known White Patches since the mid-1980s and I was struck by how she was always babysitting young calves. She was the eternal aunt. White Patches was routinely in charge of at least two or three calves at a time. When not babysitting, she hung out with the boys. But a strange thing happened by the end of the summer. White Patches, the reliable and consistent aunt of the community, was pregnant and suddenly she was taking the role of a calf, swimming underneath her protégé Diamond! Diamond rounded White Patches up as she swam excitedly. It was a complete role reversal. White Patches stopped making her distinct and well-known signature whistle (which perhaps tells us something about the function of such whistles). Typically mothers use their signature whistles to round up their calves. White Patches had routinely used her whistle to round up her young charges. Perhaps her role as aunt and babysitter was now over and she no longer had a use for her signature whistle. Whatever the reason, the change could not have been more dramatic. It was the only time I have observed the complete cessation of a signature whistle by a female dolphin. I thought, "It will be interesting to see how White Patches does with her first pregnancy," but we had already seen this strange and unexplainable change in her behavior. Pregnancy is a risky time for females. The old adult males often escort these mothers-to-be, probably for protection. Occasionally when we return the next spring the females who were pregnant the previous fall have no calves. Sometimes we never see the female again.

I remember talking to my colleague, Dr. Randy Wells, who has been studying wild dolphins in Sarasota, Florida, for more than thirty years. Randy has also documented nonreproducing females in his group and describes a similar function and reproductive status of some of his females. Personally, I like to call these dolphins "career females" since they have roles in their society other than motherhood. It could be that they are unable to physically reproduce or to carry a fetus to term. It may be that a diverse range of personalities serves a very clear function in a complex society. A number of animal studies—on primates, horses, and even octopuses—show how personality and individuality function in their respective societies. It is these complex roles in other animals and our observation of the babysitting role among childless adult female dolphins that drove my colleague Christine Johnson and I to spearhead a workshop in 2005 and publish a series of papers on comparative cognition between primates and dolphins. There is much more to complexity in a society than numbers and patterns; diverse personalities may be paramount for a healthy, functioning society. I, myself, don't have children, having never had the desire for that lifestyle. I also felt that I did not want to add to the overpopulation of humans on the earth. This choice in my life has allowed me to do regular, long-term fieldwork for more than twenty-five years. But I sympathize with other female colleagues who have struggled with balancing their desire for children and fieldwork. Most give up their fieldwork and focus on supervising their graduate students and analyzing their data.

Reproductive diversity adds to a society's resilience. In the case of dolphins, nonreproductive females allow a foraging mother a break from supervising her calf. I wonder if these roles are genetically determined or are learned behaviors. Perhaps these females are like the rogues of human society, immigrating occasionally to other dolphin groups and learning social rules that are different from their natal group but that might be crucial to survival down the road. Biodiversity is the key to biological survival and social diversity may be the key to social survival. The social ecology of a society might be as important as its environment; it is at least as diverse.

This summer there were also clear examples of the development of male coalitions. Occasionally in the field there are perfect behavioral observations and by that I mean that I knew all the players, the ages, the relationships, and the behavioral activity provided a missing piece to the puzzle. One such encounter happened in mid-July, leaving me with a keen sense of the process of learning how to be a male dolphin. Luna and her son Latitude were in tight formation close to each other while two males pursued them intensely. Punchy approached his male friend, Big Wave, inverted and then copulated with him. Luna chased Latitude inverted, usually a sign of displeasure and impending discipline from a mother. Suddenly Latitude joined the two older males and they began to synchronize their surfacing and breathing, much as an older male coalition would do in a heated fight. Latitude took protective shelter with the older males while his mother chased him and the older males smoothly incorporated him into their rhythm. Finally Luna chased Latitude out of their protective custody and he swam cooperatively with his mother, pectoral flipper on her flank. But instead of disciplining Latitude, Luna turned sideways allowing Latitude to rub his pectoral flipper against her genital area (called a pec to genital rub), a peacekeeping gesture. Punchy then encouraged Latitude's foraging for small fish on the bottom. This encounter had the elements of courtship behaviors, modeling of copulation, and the integration of the young male calf into a functioning male coalition. I also suspected that this was a strategy of the older males, although Punchy and Big Wave were only young adults, to gain access to a female by investing in their young. Richard Connor has described this potential secondary strategy for female access by older male bottlenose dolphins in Shark Bay, Australia. And it makes sense. In the Bahamas, young adult males (nine to fifteen years old) spend much of their time fighting each other, fighting other male coalitions, and displaying in front of the females. It is more likely they are practicing, building up relationships and rank, for the future. No one knows when Atlantic spotted dolphin males become sexually mature. In pantropical spotted dolphins, males must reach the fused age class, probably fifteen to twenty years or more, to be sexually mature. Our genetic work in the future would show that indeed it was the old males, around thirty or thirty-five years of age, who were primarily siring the offspring. Was this because of their age of maturity or other strategies for gaining access to females? Do old males invest in an actual relationship with a reproducing female so that when the time comes she chooses them? I suspected there were many different strategies for males to get access to females.

Developmental years for males were fraught with displays of subordination, dominance, and competing alliances. Some males found themselves on the subordinate end of the stick when surrounded by older males. This summer Geo, only three years old, was on the bottom of the totem pole. Through the summer we repeatedly saw Geo aggressively chased by other males who held him down on the bottom, buzzed him, and copulated with him. The mottled males chased Gemini, Geo's mother, away during this activity. Apparently this was an issue among the boys. During aggressive interactions with male bottlenose dolphins, the youngest of the male spotted dolphins is often the target of their attention. Being low on the totem pole was not looking like much fun and it was Geo's lot this year.

This summer I started noticing subtle but important things about male development and how they learned from one another. Latitude, Navel, and Poindexter were practicing open-mouth play and engaged in head-to-head and open-mouth swimming, under the watchful eyes of the adult males nearby. Occasionally Dash or Gray Scar, the younger of the adult males, intervened and then went back to receive a pec rub from Knuckles, the eldest. Could this be a reward for intervening the correct way? Luna, Latitude's mom, occasionally tried to intervene, but the mottled males chased her away, as if to allow the young males to continue. Was I seeing Knuckles, the oldest male, train the younger adult males, Dash and Gray Scar, how to intervene with younger males Navel and Poindexter? Boy, was it complicated! Even the mother, in this case Luna, could not override the social practice under way; in any mammal society that is saying something.

Mothers with male offspring often spent time with each other, as did mothers with daughters. I wondered if the mothers associated for the purpose of having their young males exposed to each other or the daughters exposed to babysitting rules. Could it be that females are changing their associations based on the sex of their calves? Although it never seemed statistically clear from looking at this dynamic, it sure appeared during these encounters that it was certainly an advantage to having your playmates be the same sex.

By late August we monitored the all-too-familiar hurricane patterns. Tropical Storm Cindy was two hundred miles southeast of Puerto Rico, Tropical Storm Barney hovered nearby, and Hurricane Emily was on the way. As we nervously tracked the storms they suddenly veered away as simply as the cirrus clouds above. We watched the shuttle *Discovery* launch from Cape Canaveral, star-bound explorers parting the once-threatening clouds to destinations unknown. The ocean and space: one wet and warm, the other dry and cold, and both vast, unexplored, and rich with discovery.

By the fall I was back on land, finishing my Ph.D., and getting ready for a marine mammal conference in Galveston. I had become an adjunct professor at Florida Atlantic University and begun recruiting my own graduate students for the project. There was no lack of projects off-season since the fieldwork is only a part of the scientific process. It was on land where there is time, and computer power, to analyze patterns and test theories. But it certainly wasn't as much fun!

~ 5 ~

Interspecies Interactions: Spotted and Bottlenose Dolphins

The network of relationships is the fiber that makes up elephant society.

—CYNTHIA MOSS, ELEPHANT MEMORIES

1994—Stubby's Revenge

Returning again this spring to our regular study site was like a breath of fresh air. My emotional transition from the boundary of land to the open sea is mirrored by the cobalt blue of the Gulf Stream as it changes to a lighter shade of blue, then aqua, then light green as we approach the shallow sandbank. My good friend and artist Pat Weyer described the colors to me from a glass artist label of colors: Reichenbach 55 grades to R26, to R24, and finally Zimmerman or Z110. It is eye candy at its finest and one I have tried to capture in pictures, but today it resides only in the moment. This first month of May brings not only a lunar eclipse, but also a solar eclipse, reminders of our small place in the universe, as we chug along in our small floating home on a very large sea.

By now I had worked every summer, for nine years, in the same area studying the now well-known community of spotted dolphins. I had also documented the regular presence of a second species of dolphin, bottlenose dolphins, in the area. In my early years, when mine was about the only vessel out in this area, I spent most of my time anchored. My logic was to just be in the area, be quiet, and be available. Gradually, and carefully, the spotted dolphins came by our boat. We slipped in the water, took photographs or video, sexed the dolphins, and tried to behave ourselves. The first few underwater encounters with bottlenose dolphins were during our interactions with the spotted dolphins. It's likely that the bottlenose dolphins observed our behavior with the spotted dolphins and got comfortable over time. But even today there are only a few of the resident bottlenose dolphins that are curious about us and tolerate our presence in the water.

Yosemite was the first bottlenose dolphin to break the barrier. After years of getting to know the spotted dolphin society, Yosemite, named for her half-dome-shaped dorsal fin, gracefully reached across the interspecies divide. One day, as we watched a group of three bottlenose dolphins, Yosemite broke away from the other two and circled around us vocalizing and swimming rapidly. Then she swam toward the boat and up to the gate where we humans entered the water and took a good look. After satisfying her curiosity about her human toys Yosemite went down to the bottom and pulled up a piece of seagrass that she quickly offered to us by dropping it in the mid-water column. This is a game I had come to expect from a friendly spotted dolphin but I quickly seized the moment and grabbed the seagrass with an eager offer to play. Later that week Yosemite returned to our anchored boat early one morning and began tail-slapping. As we had found with the spotted dolphins, tail-slapping is basically a signal for attention, whether it is to get the boat moving for a much-desired bow ride, a signal inviting us in the water, or the babysitter's note to call a rambunctious calf back to safety. We joined Yosemite and found three other bottlenose dolphins awaiting us. After a bit of synchronized swimming and bottom foraging this group of bottlenose dolphins lazily swam off in a spread formation. Not bad for a second date.

In the early days of my research the dolphins often came to our anchored research vessel. Most of the time this happened in a few areas where we anchor; the dolphin highways I call them. One sweltering summer day, as is often the case in peak summer months with little breeze, we were diving in off our vessel to cool our brains. Suddenly a young spotted dolphin appeared and stationed himself directly under the gate from which we were diving. Instead of grabbing cameras I encouraged everyone to keep diving. The dolphin was fascinated. He gazed up from below in anticipation of the next human and then as the diver entered the water, the dolphin carefully followed every movement of this hairy aquatic creature. Catching a human off guard was probably just fun for the dolphins, but these moments served to develop rapport with them. Sometimes data gathering took a second seat to the relationship. In this case I felt it important to show something to this dolphin that had so eagerly and clearly initiated interest. Sometimes we humans interacted with each other in the water, holding hands or playing games, to show the dolphins a bit about *Homo sapiens*. From the dolphins' perspective we were creatures who showed up seasonally, swam in the water, and oriented to them with funny metallic objects like cameras and video. The dolphins never saw us eating, mating, or sleeping (at least that I am aware of). I admit once I tried to eat a raw flounder I had caught on the bottom in front of a dolphin, to try to share the analogy of eating food. I had dreamed of building a large catamaran, with see-through hulls, so the dolphins could come by and observe us cooking, eating, and playing in our above water world. The relationship, etiquette, and rapport building with the dolphins were critical to my research as were the relationships formed by Jane Goodall with her chimpanzee families, Dian Fossey with gorillas, and Cynthia Moss with wild elephants. I needed this species' trust, and interest, to observe their underwater world on a regular basis.

In the 1960s and 1970s bottlenose dolphins were captured in the Bahamas for captive facilities in the United States. It would not be surprising, from what we know about the mental faculties of dolphins, that these animals would have long-term memories and shared experiences of being chased, roped, or having their offspring and friends captured by humans in boats. Researchers that temporarily capture dolphins for health monitoring, or tuna boats chasing schools of dolphins to set nets around them, have to alternate boat engines and boat types since dolphins recognize the sounds associated with these human activities and develop their own strategies for escape. This should not surprise us, since even prairie dogs recognize and communicate to each other the presences of humans with guns and without guns, individuals walking toward them, and other complicated labels of their world (via the great work of C. Slobodchikoff and colleagues). We will probably discover that many species are smarter, and more complex, than we have noticed.

Although bottlenose dolphins are not the main species of focus in my work we have always photographed their dorsal fins and recorded their behavior. In 1993 I received a call from my old mentor Dr. Bruce Mate at Oregon State University. He had a student looking for a dolphin project and was wondering if I had any ideas. Kelly Rossbach was our first student to focus on the bottlenose dolphins. She was as ferocious about her fieldwork as the mosquitoes on her land-based field site were about her, and we lovingly called her field site on the island the "bite site." The marina at West End had allowed Kelly to pitch a tent on the sand for her study of the bottlenose dolphins around the island. Half of the time she was based on land and used her small inflatable boat to track dolphins, the other half she was on board our research vessel, *R/V Stenella,* to work on bottlenose dolphins farther offshore. I remember taking Kelly out in the inflatable and showing her how to use the anchor. We talked of squalls and thunderheads and ways to read the weather. I admit I was a bit nervous. She was my first student, albeit jointly with Bruce, but I was her field advisor. And if there was one thing I knew, it was that the weather out here was dynamic, dramatic, and dangerous. Storms and winds could catch you off guard. Every year a fishing boat or two from West End was lost at sea, caught in summer squalls too far offshore. We decided to make eighteen miles the limit for her boat from West End but only in good weather and with radio contact. When Kelly was on our boat she went out with a jug of water, a radio, and a radar-reflecting device in case of trouble. Luckily Kelly was careful and prudent about her fieldwork, and she was able to follow groups of bottlenose, photograph their dorsal fins, and occasionally jump in the water to see what they were up to.

One day our research vessel happened to be in port so I took advantage of the time to join Kelly in scouting out the south side of the island. Before development of the western side of the island into expensive homes and a fancy hotel, the south side was full of pine trees and secluded coves, many of which the dolphins used on occasion. Today the dolphins were in a small cove off West End and we watched as two male bottlenose dolphins mounted a young adult female spotted dolphin and then the same males copulated with White Spot, a male spotted dolphin. This was one of those rare moments that we were able to follow a group of dolphins around the tip of the island to the east toward Freeport. Divers from UNEXSO had given me video of Stubby and other dolphins occasionally sighted off Freeport during their dives, but this was the first time we had seen it ourselves. It is often the case in fieldwork that we are the ones limited by how much area we can cover. It's a big ocean, and just because we don't see a dolphin one season doesn't mean they aren't there. It reminds me of the days when marine mammal field guides said that blue whales don't breach. Well, apparently blue whales didn't read the book because they do indeed breach. Kelly was an exceptional student, winning an award for her first poster at an international conference. But fieldwork is intense and science is for those that relish long-term gratification. Kelly eventually moved back to the Midwest with her new husband to start a family, but her work provided the baseline for our continuing studies of bottlenose dolphins in the Bahamas.

For years I had an eye out for an example of the "big bang," which in the dolphin research world refers not to the creation of the universe but to a theory by the late Dr. Ken Norris, who postulated that a sound can be so intense (up to 220 decibels) it could "stun" a fish, making it easier to catch. Many of us in the field were looking to observe such behavior, but in the end we observed not a bang but a buzz as the spotted dolphins scared up razorfish from the bottom. The closest thing I heard to a big bang was when a dolphin either hit another dolphin with its tail, or the tail moved underwater so fast that it cavitated and created a banglike sound. Dr. Ken Marten and I measured these bodily sounds and compared them to bangs that have been recorded in captivity and they were clearly different. We published our thoughts about types of sounds used to catch fish including the effect of the buzz on buried fish. Years later researchers in Hawaii tried to test our observations by using some artificial tanks, the wrong species of fish, and environmental conditions that were nothing close to fish living under the sand. Their conclusions were that it would not be possible to stun a fish with a buzz or a bang. The paper is an example of what not to do when trying to test a theory; if you can't replicate conditions closer to the natural environment in question, in this case fish hiding under the sand, your methodology as well as your conclusions will be flawed. What they did show is that you can't stun a fish in the water column in a tank with the sounds they projected, but that is all the study showed. And I continued to watch dolphins in the Bahamas stun fish out of the sand for their dinner with a buzz, not a bang.

Bottlenose dolphins enlist many different strategies for finding fish and the creation of a vortex underwater is one of the most amazing ones I have witnessed. For years I watched the bottlenose dolphins scan the bottom, occasionally hitting their flukes or pectoral fins in the sand as they skimmed the bottom. Lucky enough to be videotaping on the bottom, I watched with amazement as this tap on the sand turned into a swirling vortex, an underwater tornado. The sand tornado slowly moved over to a hole and hovered there as the dolphin turned back and dug his rostrum in the sand where the tornado was hovering. What had I just seen? Were the dolphins purposely hitting the sand to create a vortex? The only reason I saw it was because the vortex pulled up some sand and took a "visual" form. How many times had the dolphins created something like this and we had not "seen" it? Later I talked to a vortex physicist who told me that vortices are attracted to low-pressure areas, like holes. The dolphins had created a tool to find a fish hole. What else do they do? Could they be creating shapes in the water, invisible to us, but useful to them? In captivity dolphins create bubble rings and use them as toys.

They create air rings by streaking by quickly and throwing off a swirl of water from their dorsal fin or pectoral fin. They are invisible to us until the dolphin injects air into the ring. They are masters of the water. The use of this underwater tornado of sand to find a fish hole was one of the most amazing things I had ever seen, but it looked pretty routine for the dolphins.

Dolphins often scanned from the surface of the water down to the sand to target fish. Today I watched a bottlenose dolphin hover at the surface as she rotated her head, scanning the bottom forty feet down. Underneath the mother, her calf rotated his head mimicking his mother in perfect synchrony. Then the mother dove down to the bottom and quickly pulled out a fish. I hadn't heard one click, or sound, to indicate they were using their sonar, although if they were using ultrasonic clicks (above human hearing) I wouldn't hear them. I knew that high-frequency sounds were very directional, but usually I heard the low and audible parts of their echolocation during normal hunting behavior. Could they be listening for fish under the sand and not actively producing sounds? I could only think how much work we had to do out here and how I needed some high-frequency recording equipment.

As the summer went by we also saw bottlenose dolphins digging in squirrelfish holes. I watched a group of bottlenose dolphins and nurse sharks tuck themselves under a shallow ledge and pull out fish after fish along with an occasional lobster. No wonder the bottlenose rostrums were so scratched up! They dug under reefs, in shallow holes, and fought aggressively with each other, and they still managed to scratch their bodies on the sponges and soft coral as they meandered away after their feast.

During the summer Kim Parsons, a young student in the Abacos on the eastern side of Little Bahama Bank, contacted us. She was wondering if we could collect some fecal samples from the bottlenose dolphins. She was developing a protocol to extract DNA out of fecal material and she needed samples from our area. I knew that primate fecal material was used to obtain DNA, so I was thrilled that someone was taking a crack at doing the same with dolphin feces. Although many of my colleagues, both with larger whales and various species of dolphin, used biopsy techniques involving shooting a dart and taking a plug of skin and blubber, we depended on the good graces of our dolphins to tolerate us in the water, so I had been waiting for a benign technique. We set to work getting samples from the bottlenose dolphins and simultaneously collected samples from individual spotted dolphins. As we swam through clouds and clumps of fecal material underwater we scooped up material in a small sample jar. It's not as easy as it sounds. First you have to make sure you know who defecated, since we were interested in tracking individual lineages and paternity. It's a good thing that humans can't smell underwater because the process is not a pleasant one. In my younger days I had scooped up harbor seal fecal material to analyze their diet, but in the water we were immersed in their almost liquid defecation much of the time. I called my mostly female graduate team the "femme fecales" and we proceeded to verify, jar by jar, which individual had defecated, and we began building our collection of spotted dolphins' genetic samples.

When Kim began comparing DNA between the bottlenose dolphins in her area and our area an amazing thing happened. We discovered that some of our bottlenose dolphins on the west side of the sandbank were occasional visitors to the east end of the sandbank, a hundred miles away, and vice versa. We confirmed this again through photo-identification of their dorsal fins. We had always suspected that it was possible that both these dolphin species had larger ranges than our study area. Since our project uses only benign research techniques, our data come from dolphin sightings and photographic confirmation of identification. Now both photo-identification and genetics confirmed that some bottlenose dolphins crossed the one-hundred-mile stretch of sandbank (or went around in the deep water). What was even more interesting was that the female bottlenose dolphins were moving around the larger sandbank, not the males. It's a pretty well-accepted fact that males, at least in mammals, are the ones to roam around spreading their genes. Female-mediated gene flow was virtually unheard of in the mammalian world. Here we had female bottlenose dolphins roaming around the sandbank, perhaps an artifact of the unique environment of the Bahamian seascape. Over the years I have done winter trips all around Little Bahama Bank, including the Abacos, the south side to Freeport, across the sandbank west to east, and around the bank in deep water. Never once had we seen a spotted dolphin farther into the shallows eastward than about ten miles. The bottlenose dolphins were there, in the shallows, near the flats that seemed to be their niche and probably what allows the two hundred individuals of both species to live together; they don't have to compete for food. An adult bottlenose dolphin on the banks is three feet larger than a spotted dolphin and requires more calories per day than the average spotted dolphin, which is probably why the bottlenose dolphins do more foraging and less socializing than the spotted dolphins. A human society with plentiful resources of food and shelter has more social time to develop art and culture. I suspect the spotted dolphins are more like such a human culture with many resources, spending only a part of their day feeding with time left to play and socialize; another lucky break for our work.

One of the most surprising aspects of my work in the Bahamas was the gradual realization of the intimate interspecies social relationships between these two species of dolphin. I remember the first time I observed these two species together. In 1985 I got a hazy glimpse through murky water of a bottlenose dolphin in a head-to-head posture with some spotted dolphins. I remember thinking, "Too bad the video will be murky, I will probably never see this again." But I was wrong. Of all our encounters, we observed spotted and bottlenose dolphins together 15 percent of the time. Much of this time the two species were traveling, playing, or socializing together. More than 30 percent of the time the males were fighting, which takes the form of the physically dominant male bottlenose bullying and mounting the young male spotted dolphins. Usually outnumbered, it was dangerous for young male spotted dolphins to try to escape, since the bottlenose dolphins can inflict even more damage if they try.

To complicate things even further, interspecies mating attempts were not uncommon in the Bahamas. Occasionally male bottlenose dolphins chased and succeeded in mating with female spotted dolphins, despite the blocks and aggression from the male spotted dolphin coalitions. One summer, Hook, a female spotted dolphin, was the intense focus of male spotted and bottlenose dolphins, and today forty-two spotted and nine bottlenose dolphins were here. The male bottlenose dolphins held down and mounted Flash, a young male spotted dolphin. Flash and Dash competed with each other for access to Hook but also competed with Liney and Slice, two other male spotted dolphins. As Flash approached Hook inverted to solicit mating, she tail-slapped him and darted hastily away. Throughout this encounter both the male bottlenose and male spotted dolphins attempted time after time to mate with Hook. As we had seen all summer, Hook was a focal (probably in estrus) female getting much attention from the males throughout our field season, but it seemed that her signals were not species specific.

Over the years we saw male spotted dolphins distracting male bottlenose dolphins from trying to mate with the female spotted dolphins. In one such encounter we were swimming back to our boat when we heard a cacophony of squawks and whistles typical of interspecies group vocalizations. As I watched, a spotted male coalition swam by with Trimy, a young adult female spotted dolphin, with two large male bottlenose dolphins following closely. The male bottlenose dolphins began chasing Trimy, but were not yet fully aroused. While the male spotted coalition huddled, one of the male bottlenose dolphins got an erection and swam toward Trimy. Navel, the youngest of the male coalition, went over to the male bottlenose dolphin and solicited his attention. As Trimy escaped back to the male spotted coalition the male bottlenose dolphins chased and buzzed her with out-of-water leaps and tail whacks. The male bottlenose broke off from his second chase of Trimy and expelled a large bubble at Navel. Successful at running interference for Trimy, Navel was the target of the male bottlenose. Perhaps Navel knew he could get the attention of the male bottlenose dolphin to distract him from Trimy. Perhaps he wanted the attention of the bottlenose dolphin. A month later we saw Trimy pursued again by a male bottlenose dolphin. In this case

Trimy and the bottlenose male were mating with unusually intense enthusiasm. Pimento, another female spotted dolphin, was also the focus of the male bottlenose dolphins' attention but with less vigorous mating attempts. Throughout this event, Duet, a young male spotted dolphin, tried his courtship skills with Trimy and managed to copulate a few times. Then the male bottlenose dolphin took on Duet, challenging him by passing perpendicular in front of him, and then copulated energetically with Duet to complete his activities.

Even more rare is the observation of male spotted dolphins mating, or attempting copulation, with female bottlenose dolphins, although we do see dominance directed toward female bottlenose dolphins. As I watched in surprise, Liney and Slice, two male spotted dolphins, were focused on mounting one of the female bottlenose dolphins. They herded, inverted chased, and side-mounted the female bottlenose dolphin exactly as a male bottlenose would do to them (male spotted dolphins mate belly to belly, not from a side position). Although there were indications of aggression, including tail thwacks, head-to-heads, and open-mouth displays, it didn't have the quality of highly escalated aggression. It seemed more an exercise in dominance by the male spotted dolphins.

Just when I thought I had seen it all between these two species, I witnessed two female bottlenose dolphins in their solicitation of two young male spotted dolphins. While the female bottlenose dolphins actively solicited the attention of the male juveniles spotted dolphins, these young males went limp and passive, as they do with male bottlenose aggression. One of the female bottlenose dolphins seductively wagged her flukes in front of the young male spotted and then rolled so that his flukes were scratching her head. The female then began to mount him and as she held him between her pectoral flippers she thrust her body against him. Such female solicitation of sex from a young spotted dolphin was unique, and read somewhat like a bad romance novel, but it was apparently part of their complex relationship.

Bottlenose dolphins have mated, and produced hybrid offspring, with more than thirteen different species in captivity. In at least one case, where a bottlenose dolphin mated with a false killer whale, the hybrid dolphin was fertile and went on to reproduce herself. For species in the wild, the only way to remain reproductively separated (when in the same geographic area and physically able to reproduce) is by a behavioral reproductive isolation mechanism. During my years of observations of interspecies interaction I have witnessed male bottlenose dolphins attempting to mate with female spotted dolphins. But male spotted dolphins often interceded and fought off this attempted interspecies mating. And I suspected that one of the main functions of male spotted dolphin coalitions, and the spotted dolphins aggressive interactions with male bottlenose dolphins, was to prevent spotted females from interbreeding with the bottlenose dolphins.

Interestingly, a very small subset of bottlenose dolphins in this area have black spots on their ventral side and are friendlier than other bottlenose dolphins. This leads to the thought that these "spotted" bottlenose are perhaps hybrids (which do often show both morphological and behavioral traits of each species), something we hope to look at with their DNA in the future. This may not be surprising since spots are thought to function as a camouflage and resemble sunlight streaming through the water and hitting light-colored sand. Bottlenose dolphins in the Indian Ocean do have spots on their ventral side. But if these spots are evolutionarily advantageous, why don't all the bottlenose dolphins in this area of the Bahamas have spots? Hybrid baboons are known to share behavioral and physical traits of both species. In East Africa, Karen Strier and colleagues describe how behavioral traits are shared in the cross-migration and hybridization of Hamadryas baboons and Olive baboons. Could it be that spotted and bottlenose dolphins have been interbreeding here on a regular basis and that is why they sometimes share morphological and behavioral traits? Dolphins are karyotypically similar, with the same number of chromosomes, making them better able to successfully crossbreed than other animals. They also have great mobility in the open ocean and are therefore able to encounter each other more than many terrestrial species that are restricted by mountain ranges and continents.

Could hybridization of spotted and bottlenose dolphins act as a potential speciation mechanism? Fertile hybrid offspring between blue whales and fin whales have been identified from whaling records. But a hybrid dolphin would have to breed with another hybrid to create a hybrid population, or to affect population genetics significantly. Sonja Wolters and Klaus Zuberbuhler describe two species of monkeys in Africa, the Diana monkey and Campbell's monkey, where two adjacent communities have interbred so much over the years that they have produced a hybrid border community.

In the terrestrial world most interspecies mating encounters are one-way, with the physically larger species having the dominant advantage. Pamela Willis and colleagues have documented hybridization activity between two species of porpoise in the Puget Sound, Seattle, area of the United States. Dall's porpoise sometimes mate with Harbor porpoise and produce hybrids. These mating encounters are one-way; the larger Dall's males mate with the smaller harbor porpoise females. Therefore the hybrids have a female harbor porpoise as their mother and a Dall's porpoise as their father. The individual, or species, that is dominant has greater control over the outcome, either by pure size or rank. The subordinate has less opportunity to shape the outcome of an encounter. Our own work with spotted dolphins and bottlenose dolphins in the Bahamas shows a strong tendency toward bottlenose males mating with spotted females. Yet the earlier incident shows that the female bottlenose dolphins seek out and successfully copulate with male spotted dolphins on occasion, although the males in that case may be sexually immature and incapable of reproducing. Our only observation of a hybrid in the Bahamas comes from a secondary study site in Bimini, where the mother was a spotted dolphin and the father a bottlenose dolphin. Perhaps in the future genetics will provide the information to verify whether hybrids exist or not in this area.

Like other mammals, dolphins have normal aggressive behavior within their species to settle conflicts. One of the most aggressive encounters I observed with the spotted dolphins took place while two southern males,

Sickle and Quotes, were chasing a northern female, Paint, in front of the northern male spotted dolphins. As the southern males tried to copulate with Paint, she swam off and then returned to confront them head-to-head, fending off their advances with tail slaps. Meanwhile the old fused males from the north were huddled together and began to gang up on Sickle, doing stationary head-to-heads, open mouths, screams, squawks, bubble (donut-shaped) rings, and every other aggressive signal in their repertoire as they tried to hold Sickle underwater and cut off his access to the surface to breathe. It was the most violent behavior I had observed among male spotted dolphins and as might be expected it was over a female.

Male-to-male sexual behavior between the two species is fairly ritualized. Typically a small group of male bottlenose dolphins approaches a small group of male spotted dolphins, with at least one spotted dolphin on the younger side of adulthood. The bottlenose males coordinate their activity, approach and mount a spotted dolphin, usually the youngest one, with full erections. Sometimes this is preceded by chasing and aggressive behavior before such arousal is observed. When a male bottlenose has targeted a spotted dolphin they stay focused on that individual. The spotted dolphin, unless he has a helpful male coalition, goes limp and passive, sinking vertically and allowing himself to be rolled around like a log in the water, or he swims slowly behind the larger bottlenose that tail-slaps him on the head. The consequences for taking on a large, aggressive bottlenose dolphin without your male coalition nearby is severe. The whole dynamic of the interaction changes when a larger group of male spotted dolphins enters the picture. This is why it is so important to document the entire sequence of behavior: who was there in the beginning, how many of each species, what were they doing; did the ratio of the species change during the encounter, and if so, what ensued?

If there is one story that describes the intensity of the interactions between male spotted dolphins and male bottlenose dolphins, it's the incident I call "Stubby's revenge." Every once in a while you stumble on to something that gives great insight into dolphin life, and it happened during a film shoot with Robin Williams, for the series *In the Wild.* In the winter of 1994 I received a visit from a film company based in the U.K. doing a series on celebrities and animals. They picked a celebrity, found out which animal they were interested in, and took them around the world to see their animal of interest. Anthony Hopkins had done a program with lions, Julia Roberts with orangutans, and now they had someone interested in dolphins, and could we help? Well, my field season was pretty booked up already with various project personnel and volunteers, but it so happened that a board member had just canceled a trip in June. I wondered who the celebrity was and if they would be a good spokesperson for the dolphins. I had tried, over the years, to be selective about the film crews we brought out with us. If they weren't going to spend enough time to do service to the dolphins, I wasn't interested. Everybody and his grandmother wanted underwater footage of these now known wild dolphins, but I had made it a point not to serve as a boat for photographers or filmmakers, unless it was for a legitimate natural history film. Unfortunately photographers often have a different agenda than our research and the two goals often meet head on. Anyway, I thought Leonard Nimoy would be good, or Jodie Foster. When the producer finally told me it was Robin Williams, I thought, "Yes, he could definitely tell the story." So in the summer of 1994, Robin Williams joined us for what would be both a great dolphin trip and four days of hysterical laughter as well as serious conversation. Robin himself would be a witness to "Stubby's revenge."

As we cruised along looking for dolphins we found a large mixed-species group. When we entered the water the most amazing scene was playing out. Stubby and many of his male associates were there in a large, huddled mass with four large male bottlenose dolphins. As we watched the scene unfold, two aroused male bottlenose dolphins joined Stubby as he broke away from his spotted group. The whites of Stubby's eyes told me everything: Stubby was being bullied and harassed by two large, sexually aroused male bottlenose dolphins, Sly, with a deformed left pectoral fin, and Max. I had seen the consequences of a male spotted dolphin trying to escape the larger, physically dominant bottlenose dolphins. Three feet larger and with a temperament to match, the bottlenose dolphins could easily physically dominate a spotted dolphin. But today was different. Stubby was in full force and flanked by more than a dozen male spotted associates, a supercoalition of dolphins including his own long-term coalition Knuckles, Punchy, and Big Wave. Stubby wasn't going to take this kind of interspecies domination, at least not today. This supercoalition of male spotted dolphins appeared to be out for revenge.

The bottlenose dolphins mounted Stubby and attempted copulation when suddenly Stubby broke from the aroused bottlenose dolphins and rejoined his spotted dolphin group. Then in a display of what I can only call coordinated effort, Stubby led his group in a chase after the bottlenose dolphins, specifically focusing on Sly. Whenever the spotted dolphin coalition caught a glimpse of Sly on the bottom, they mounted a charge toward him with Stubby leading the way, pushing and biting Sly until he fled from the bottom. Sometimes the water surface boiled and broke with energetic leaps of Sly escaping the supercoalition's heated attack. The male spotted coalition grouped and regrouped, swimming tightly next to one another, pec rubbing and squeaking until they calmed down. Even for dolphins, this encounter seemed physically exhausting.

Sly's strategy was interesting. When the male spotted dolphins chased him around, Sly dove down to the bottom and began to, what looked like, crater feed. I got the strong sense that this was actually a displacement behavior by Sly, something to do to disengage from the other activity and distract the coalition. As Sly settled into his digging, the male spotted dolphins began regrouping, frantically vocalizing and touching one another in what looked like something akin to a football huddle. As an observer in the water, I felt the excitement, too, and I found myself merging with the group and vocalizing myself through my snorkel, "Way to go, let's get him"; perhaps a real-time example of interspecies social contagion; I felt like I was one of the group. I moved my bent elbows to communicate my willingness to pec rub in excitement (even though we have a policy not to touch the dolphins). The energy in the water was electric, the noises deafening.

Stubby was with Punchy, Rivet, Groucho, and Knuckles while Flash and Dash were tight with Big Wave. They joined as one larger supercoalition and then turned, sometimes 180 degrees, and targeted Sly. The bottlenose energetically leaped out of the water as the spotted dolphins continued their rally. Sly stayed in the area, even though the male spotted dolphins swam around in tight groups with open mouths, huddling closely. The contagious sense of escalation, combined with the displacement foraging behavior of the bottlenose dolphin, was something to be experienced. My blood pressure was up, I was agitated, and I was with the coalition, body and spirit.

That day spotted dolphins outnumbered bottlenose dolphins. Chris Johnson and I had calculated that a ratio of six to one of spotted dolphins to bottlenose dolphins was favorable for turning the tables to the spotted dolphins' advantage. This scene played out over and over again. Whenever the group saw Sly on the bottom, they mounted a charge toward him, fought for a while, and leaped out of the water in aerial displays that looked akin to play when viewed at the surface, but were usually representative of escape from a chase below. The male spotted dolphins regrouped and swam around with bubbles streaming from their blowholes and wide, open mouths until they calmed down. Now, trying to document this event while swimming around with a video camera in strong current can be quite exhausting. I finally realized that I could remain at the surface above Sly, watch him while the spotted males regrouped, and sure enough they would find Sly . . . and there I would be at the center of the action. I was in the middle of it all and thought I was about to be rammed by Sly during the heat of the moment, but instead he just leaped over my head and continued on. This went on for four hours and possibly longer, and in the end, as so often happens during encounters, we could no longer keep up with the dolphins.

This event clearly demonstrated the power of the male spotted coalitions especially when in a supercoalition. But what made it even more extraordinary was the previous day I had observed Stubby with a mother-calf group. Here, too, he was approached and mounted by Sly and another male bottlenose dolphin. With only one other male partner there, Flash, a mature spotted, Stubby had no group defense. Sly and his male friend were ruthless, mounting and pushing Stubby around, simultaneously aroused and acting in an aggressive way. His behavior, as was typical when spotted dolphins are outnumbered by the physically larger and dominant bottlenose dolphins, was to go passive, allowing himself to be bullied by the bottlenose dolphins as they became more and more aroused. When the bottlenose dolphins left, Stubby swam off with his mother-calf group.

But today it appeared that Stubby had somehow managed to rally his group of male associates together for this rendezvous. Is it possible that Stubby registered the previous encounter as somehow unfair and retaliated against the bottlenose dolphins? Did he plan and recruit his male associates, leading them to Sly? Was the chase of this specific individual purposeful and premeditated? How would he have communicated such a message? This would imply the dolphins have the ability to plan and communicate their plans. This extraordinary sequence of events begs such a question. Or perhaps Stubby just ran across some male friends and in real time happened to run into Sly. This to me seems more unlikely than the alternative. Stubby was out for revenge and he knew how to rally help. If so, the implications are staggering, and all the more reason to decipher and decode dolphin communication signals.

Although Robin Williams was only out with us for four days (the film crew stayed longer to get other dolphin behavior), he was able to see just about every example of dolphin behavior. His first night on board I showed him video examples of underwater behavior so he could get up to speed. He told me later that he figured he would never see it all. But he did, from the moment our boat left West End we saw bottlenose dolphins feeding. Then over the next four days we saw Mugsy and Mel playing, bottlenose dolphins fighting with spotted dolphins, and he met Stubby close up. Robin was a great behavioral observer, probably stemming from his keen eye for human behavior. I would have hired him in a second had he not been a celebrity, but I never forgave him for making good imitations of the dolphin's sounds so easily; it had taken me ten years!

After studying this group of dolphins for more than twenty-five years, I recognize now that it was a spectacular streak of luck to have sighted and observed Stubby for two days in a row. Such sequential observations are rare and preciously illuminating. Normally these events might be observed over days or weeks. Sequential events like these allow previously unobserved sequences of activity to unfold in front of our eyes in real time. Such a complexity of interspecies behaviors requires complex memories and communication and perhaps intimate relationships between these two species.

I had journeyed to the Bahamas to observe the spotted dolphins, aware that they were approachable and observable underwater. In the beginning of my fieldwork I had no idea I would observe such a complex relationship between two species of dolphins. Some of the most recent genetic information, produced by Rich LeDuc and colleagues, show that Atlantic spotted dolphins are much closer to bottlenose dolphins than to others of their same genus, *Stenella*. As the bottlenose dolphins got comfortable with our human presence in the water we saw more and more of the two species playing together, socializing, or foraging. Even more amazing, the spotted dolphin females were babysitting the bottlenose calves, interspecies babysitting. Why have such a complex relationship with another species? They don't compete for food resources; the spotted dolphins utilize bottom fish on the sandbank and work the offshore edge for squid and flying fish. The bottlenose dolphins dig deep into the sand and grass beds and pull out eels and work the reefs for snapper and squirrelfish. The two species don't seem to be competing for territory, nor should they be competing for mates.

The reason for these complex and intimate interspecies interactions in the wild is not always clear. In addition to babysitting, socializing, and traveling together these two species form temporary interspecies alliances when a third-party intruder enters the scene. The first interspecies alliance I observed involved two male bottlenose dolphins that were mounting Gray Scar, a young male spotted, and repetitively tail-slapping him on the head. Gray Scar was arched and in a submissive posture typical during interspecies male dominance behavior. A second juvenile spotted joined Gray Scar and they swam together with a second bottlenose dolphin still pursuing them. Suddenly a large bottlenose dolphin, heavily infected with lobomycosis (a contagious fish fungus) approached this group and the two spotted and bottlenose dolphins immediately went into synchronous surfacing behavior. Over the years I had learned that such synchrony usually occurred during times when coordination and cooperation were needed, at least within the spotted dolphin groups. The spotted dolphins joined the bottlenose dolphins in a head-to-head face-off with this large, infected bottlenose dolphin. Such interspecies alliances in the cetacean world are difficult to observe in most field sites. Here I was observing a functional interspecies alliance between these two previously fighting species. In the future, I would observe similar alliances when a shark would enter the scene. These two species acted cooperatively with each other more often than not. It seemed more important to help each other as interdependent dolphin neighbors than to remain their own independent species.

Other times I saw evidence of the complex rules and etiquette during these mixed-species alliances. As I observed a mixed group, Navel, a male spotted dolphin, was jockeying around with the bottlenose dolphins in an open-mouth and head-to-head dance. The older male coalition of spotted dolphins hovered nearby monitoring the situation that was rapidly escalating between the two species. The monitoring spotted dolphins got tighter and tighter and more synchronized, opening their mouths and squawking wildly as the bottlenose dolphins got a little too rough with Navel. At certain points the spotted dolphins approached and disciplined Navel. Was it inappropriate for a young male like Navel to be fraternizing with the bottlenose dolphins? Were the bottlenose getting too rough with Navel and were the monitoring elders expressing their concern? Or was Navel crossing the line and getting too aggressive with the bottlenose dolphins? There seemed to be some rules of engagement at work, albeit complicated ones. Cooperation between bottlenose dolphins and spotted dolphins could be a driving force in the evolution of such a complex relationship, and it speaks strongly of culture and society rules that sometimes override species prerogatives. Perhaps the human race could learn a few things from the dolphins.

Competition, as the predominant process in evolution, has been generally accepted, although cooperation versus competition has been hotly debated for centuries. As we study and understand more complex interactions with social species, it appears that cooperation is more conducive to survival than competition. Frans de Waal, in *The Age of Empathy*, talks about the many examples of cooperation as a driving force through empathy. Marc Bekoff, in *The Emotional Lives of Animals*, echoes the same. The biological mechanism for the ability to empathize resides in a group of neurons called mirror neurons, found in higher species and activated in the process of accessing and understanding complex behavior and emotion. It allows us, and likely many other social species, to "feel" what it is like for the other.

Interspecies relationships form early in a dolphin's life. It was not uncommon to observe bottlenose dolphin calves playing with spotted dolphin calves. Interestingly, many of the developmental activities the spotted dolphin calves engaged in, such as mock fighting, chasing, and teaching, I have observed between the two species. By the time spotted dolphins reach adulthood (fifteen years or more), they may already have existing "friendships" with bottlenose dolphins in the area as they have with their own species. At a minimum they have had exposure to the communication signals and rules of their neighbors. Knowing your neighbor well is a powerful addition to the already supportive structure of a dolphin society in the wild and must lend some advantage to survival.

It turns out that if you start looking around the world there are multiple examples of mixed-species groups of dolphins. Because we work in the water we are able to observe the regular and behaviorally complex aspects of these interactions. In Hawaii, Bill Perrin and others have reported mixed groups of pantropical spotted dolphins with spinner dolphins, and colleagues in Greece have reported mixed groups of common dolphins, striped dolphins, and Rissos's dolphins. Other mixed species reports come from areas including Costa Rica, Venezuela, and Brazil. Every basic context of behavior we saw the spotted dolphins engaged in we also observed during interspecies interactions including traveling, babysitting, foraging, and playing. Groups of pregnant female spotted dolphins traveled with pregnant female bottlenose dolphins. Both species played together in games of seaweed or keep-away, males with males, females with females, and males with females. On very rare occasions the two species foraged together, but never cooperatively, and when they were digging in the bottom in the same area they didn't appear to compete.

We saw both species tightly packed and digging in the bottom in a very small area. I named this behavior "dense foraging" since the dolphins were lined up as densely as sardines in a can. The sound was that of a barbershop with a hundred electric razors going at once. The group of forty spotted dolphins wasted no time as they buzzed and immersed their rostrums deeply into the sand. Little tunny (a species of fish in the tuna family) frantically zoomed in and out trying to steal fish that the dolphins had chased out of the sand. The stunned fish emerged at the top of the sand and were easily caught by all the predators including us (for samples). I watched a lesson in table etiquette as a spotted dolphin grabbed a fish that another dolphin had chased out of the sand. I happened to be on the bottom and watched helplessly as the offending dolphin received a powerful tail thwack from the other. His head bounced off the bottom a few times. I guessed it was simply bad table manners. The stunned dolphin recovered but I wondered how often this type of behavior inflicted serious injury. Dolphins are powerful and such a blow to the head could be damaging. Even sounds, because water is a denser medium than air, in which sound travels four and a half times than it does in air, could cause damage as molecules wreak havoc on tissue.

Although not observed on a regular basis, one of the most interesting interactions between the two species is interspecies babysitting. Back in 1989 I had a brief glimpse of White Patches babysitting a young bottlenose calf during a winter trip. Years later, Venus, a maturing female spotted dolphin, babysat a bottlenose calf during an encounter on the south end of the island. In both cases the young female spotted dolphin was reaching sexual maturity and perhaps was practicing for motherhood as they did within their species. But think about it. How amazing and how trusting is it to allow another species to tend your young? It is either very reckless or very adaptive. Most of these female spotted dolphins had plenty of opportunity and already had experience babysitting young spotted calves that were their siblings or community members. So why did they also babysit young bottlenose calves? What would be the advantage? Perhaps one is simply that the youngsters of both species play together on a regular basis. One sultry day we picked up Snow, Pimento, and Little Ridgeway and her calf on the bow. Pimento was beak-genital to Snow, and Snow was flexing as if she were pregnant. Little Ridgeway's calf retrieved a piece of sea grass on the bottom while the other females rested on the sand. Five bottlenose dolphins arrived, including Yosemite and a female calf, another bottlenose mother-calf pair, and one single bottlenose dolphin. The bottlenose calves began coordinated play and chased the spotted dolphin calf around as they dug in the bottom together. The three adult bottlenose dolphins hung on the bottom while their calves frolicked with the other species, sometimes even crawling on Yosemite's head together. Barb and Trimy, two very pregnant spotted females, were nearby watching as they touched each other and rested on the bottom. It was like an interspecies birthday party, the kids played happily as the mothers lounged around enjoying a well-deserved rest. Perhaps it's that simple!

There are many interesting examples of interspecies interaction in the terrestrial world. In the Samburu Reserve in northern Kenya, a lioness adopted a baby oryx, usually a prey species. The lioness defended the young antelope from other lions and treated it like her own young (until the lioness got too weak from not eating and was unable to defend the young antelope anymore). In Kenya, a one-year-old hippopotamus, Owen, formed a bond with an old tortoise, Mzee, after the tsunami in 2004. On occasion, the spotted dolphin mothers left their little nursery group with us humans who were playing on the bow and went off to feed. Perhaps it was just a quick respite from the demands of an energetic calf as usually there was an older babysitter around. I know the mothers were not far away, perhaps satisfied in weighing the benefits and risks of leaving their precious newborns with the pleasant but aquatically challenged humans. And although I found their trust flattering, I felt it might be a bit maladaptive as our abilities to defend a young dolphin calf from harm in the water would be pathetic.

The now too-familiar late summer storms approached and both dolphins and humans could feel the effects of swells and storms. Tropical Storm Chris was rapidly building to a hurricane off the Leeward Islands but suddenly swung north toward Bermuda. As autumn equinox approached we once again said goodbye to West End, and to the dolphins, and headed to our terrestrial homes for another winter. The long rolling swells of the Gulf Stream, mere remnants of Chris's energy, were both powerful and magnificent, as we headed westward away from our interspecies adventures.

~ 6 ~

Coming Up for Air

Differences in human-animal interactions may help to explain why some studies "work" and others do not.

—Daniel Q. Estep and Suzanne Hetts, *The Inevitable Bond*

1995—Other Species on the Sandbank

Early May slammed us with wild and unusual weather for our spring start to the field season. Fifty-knot winds and cold fronts moved through Florida and the Bahamas and pounded us through May and June. Even Hurricane Alison had already come and gone by June 6, an early start to the hurricane season. Regardless of the weather we had big film plans this year. The BBC was doing a thirty-minute natural history documentary of our work. I knew both the quality of the BBC film crew and their abilities to do it right. Tom Fitz, a longtime friend, was to be in the field with us most of the season. I had met Tom in San Francisco during my graduate work. Tom went on to study underwater filmmaking, but I knew he had a good eye for dolphin behavior and he joined us for the whole field season to shoot *Dolphin Diaries*.

It was now my eleventh year in the field and I could easily recognize most of the hundred-plus individual spotted dolphins and their patterns of behavior. Familiar dolphins like Rosemole, now with her second offspring Rosepetal, were seen with her old friend and now successful mother Mugsy, who had young Mel in tow. Little Gash, with her two-year-old offspring Little Hali, was one of the "hot" females this summer, and today Liney was furiously chasing her as if his levels of testosterone were high. As some females do when rejecting a suitor, Little Gash swam rapidly away, tail-slapping him in the face as she departed. White Patches, so familiar with her distinct underbite and white spots, also made an appearance and despite being pregnant again last year, she had no calf. I couldn't help wondering if she had lost her calf or had a false pregnancy? She was back to her old tomboy ways, but still emitted only quiet little whistles.

Through the summer Tom worked diligently getting video clips of great behavior for the documentary. Little Gash gave us a spectacular sequence as Punchy, Big Wave, and Knuckles took turns and chased her inverted, stacking up underneath her for mating access, and expelled large bubbles of air in between mating attempts, all in the presence of Little Hali. Talk about exposure to social rules! Mugsy and Mel were also star players this summer. Mel had been a rambunctious and interesting calf for the last two years but now Mugsy, halfway through her new pregnancy, was ready to cut the cord. Mel had many predominant black spots for his age and it was clear that he was already on his own, hanging out with other juvenile males, as his mother got ready to give birth again.

In late June we saw Sly, the bottlenose from "Stubby's revenge," challenge a large group of male spotted dolphins including Punchy, Big Wave, and some other male young adults that were chasing Big Gash. Shorty, Ridgeway, and Pharoh, a strong, old adult male trio hovered on the periphery. As the aggression started to escalate with head-to-heads, open mouths, and jaw claps, four of the males broke off and stationed themselves head-to-head with Tom, the cameraman. Occasionally, when we don't get out of the action quick enough, the dolphins direct their signals at us, and this was the case as Tom took on the angry group of males. Eventually the males realized that Tom was no threat and they drifted away. Sometimes when I am shooting my video I suddenly realize that I may be choosing sides, aligning myself with one coalition versus the other. At least it might seem so to the dolphins. After that, Sly came into the group but was soon chased off into the distance as the rest of the spotted group rallied after him. Sly was seen porpoising away while the frantic and noisy spotted dolphins tried to catch him.

By mid-August we were again on our way up to our northern area, after days of no dolphins, when suddenly four juveniles, Baby, Rosebud, Tink, and Caroh, were at the bow. As they played and socialized in the water, Nassau and Priscilla appeared with Stubby. An example of male babysitting, Stubby hung with Rosebud pec rubbing her fluke, head, and fins. Or was this Stubby's way of gaining favor with Rosemole, the mother likely to be coming into estrus soon? Stubby traveled slowly on, tending his all-female harem of juveniles with delight.

By late August we had seen most of our regular dolphins. White Patches was here and once again pregnant (or so it appeared), and Rosebud, her companion today, was the focus of the juvenile males. As Rosebud mated with Everest, White Patches intervened, only to be confronted with a head-to-head from Rosebud's suitor. Meanwhile Stubby monitored the activity and the young males continued their coalition behavior. After all, practice does make perfect and it is so in the dolphin's society. As late summer flew by we saw the parade of very pregnant females, with their old male protectors, swim by. This early September day it was Dos and Blotches, both very pregnant, traveling with Shorty and Surge. They were very mellow and stayed in their gender-specific dyad while they traveled. All seemed normal for the end of the season.

It is rare to see any other dolphin species on the shallow sandbank other than spotted or bottlenose dolphins. Colleagues Dr. Christine Johnson and Dr. Thomas White were out with me late summer and as we cruised down the edge of the sandbank we saw what looked like a small whale. Anxious to get a close-up look we launched our small tender from the mother ship and as we approached we saw clearly that it was a small whale, actually a large dolphin by taxonomic standards; it was a false killer whale. Looking more like a pilot whale than killer whale, he was alone and full of rake marks and scars. Trying not to add to his stress we kept a reasonable distance, knowing that he was probably in the shallow water to rest and avoid the likely predators in the deep water.

A member of the dolphin family (Delphinidae), false killer whales are "group" animals, and are actually known to eat small dolphins as they escape tuna nets. A few years later we would see them while crossing the Gulf Stream, and I can tell you that I would not, voluntarily, jump in the water with a false killer whale. In the Gulf Stream they were hunting and we were in their eyesight. They observed us by spy-hopping out of the water and looked up at us from the bow. In their healthy state they are cunning, intelligent predators, whose coordinated swimming and eyeing of our boat gave me shivers as to what might happen if we fell in the water. They were hunting! Orcas, the true killer whales, show cunning and calculating displays dislodging their penguin or seal dinners from the ice floes by creating large waves to wash their prey into the water. At that moment, in the Gulf Stream, I felt for a brief second like a penguin that was being assessed by a large predator. Luckily our boat was still afloat and although I allowed some sound equipment to be lowered overboard for recording, I was not going to let any research personnel in the water, not today. I was once again made aware that we are in the food chain when we work in the water, a potent reminder of our place in the animal kingdom.

As we watched this false killer whale on the shallow sandbank it became clear he was alone and sick. Large sharks regularly patrol the edges of the sandbank. Over the years we have seen evidence, from our colleague Diane Claridge who works in the Abacos, of orcas in adjacent areas hunting many species of marine mammals. But the false killer whale floating next to our boat today would not be doing any hunting of us or of the spotted dolphins. Today he was in danger of being the hunted.

Over the years it has been hard to find our southernmost group of dolphins; their patterns and routines are not well known to us because our focus is to the north and many reefs prevent us working in the south. This year we sighted a well-marked dolphin named Triangle whose dorsal fin was much more the shape of a dorsal fin of a shark than a dolphin's (which is usually curved backwards). We had seen this dolphin every three years up in our regular northern area, but this summer we saw her with females in our southern group—a bit of a surprise, but it seemed she was "associating by reproductive status" this summer.

Whenever possible we tried to extend our range of searching for dolphins; this usually occurred during good weather stretches. For years we had heard about a group of spotted dolphins in the Abacos. Although a local dive boat operator had shared some photographs, we had never run into this group. On this fine September day we were heading home from a trip to Walker's Cay, the westernmost island of the chain of Abacos, and ran into a large group of spotted dolphins. Although they came porpoising into the boat we didn't recognize any of them (even though we got some good video). They were very friendly, whistling excitedly, and they seemed comfortable with human bodies in the water, as if they knew us. Yet to us they were a new group of spotted dolphins way up north! Although we would never see individuals from this group again, in later years we saw our southern spotted dolphins, now including Triangle, up in this same northern area. This summer was just full of surprises.

Mid-August roared in, as the summer had begun with violent weather. Hurricane Erin had forced us into port and Hurricane Felix was making its way west despite the seductive flat seas in our study area. As we crept along with our work, in between threats of Tropical Storm Gerry and Hurricane Humberto, the summer drifted into fall and another birthday at sea.

Every once in a while we have the opportunity to observe something truly unique. As we were wrapping up our BBC shoot with Tom in mid-September we were preparing to enter the water to observe the four spotted dolphins that were riding the bow. We got in and traveled slowly with the group. One of the dolphins arched and expelled a large bubble as a twelve-foot hammerhead shark swam under us. Then suddenly we started hearing loud vocalizations and a small group of bottlenose dolphins appeared. The old and apparently very wise coalition of male spotted dolphins—Shorty, Surge, Mask, and Ridgeway—kept their distance and swam slowly on the periphery as we observed the erupting violent activity. Four bottlenose dolphins were ganging up against one or two others who kept straying to the periphery. There were squawks, digital trills (apparently unique to bottlenose dolphins), loud earsplitting jaw claps, and tail whacks to the head. Many sounds the dolphins were making were "percussive" sounds, created from the strike of a fluke to a head or a rapid closure of the jaws. I swam backwards to try to get some distance from the dolphins because their sounds actually hurt my ears underwater. (Dolphins can make vocalizations over 220 decibels, which translates to a loud firecracker. In the case of body slaps and hits, they were equally as loud.) Before this escalated aggression broke out I witnessed one bottlenose dolphin hold down another bottlenose near the bottom and heard the loudest "wailing scream" I have ever heard: the closest description would be a human baby cry; it had that aural quality. In spotted dolphin society a hold-down behavior occurs in the context of discipline, usually when a mother is reprimanding an out-of-control calf. My colleague Dr. Adam Pack in Hawaii has observed such "discipline" behavior with male humpback whales underwater and seen it result in what could only be described as a whale "scream." It seems that many of these cetacean vocalizations may be universal.

I realized that this was the most intense and violent aggression I had ever observed underwater. We had very few observations of bottlenose dolphin social behavior in the water. Now, after watching this extreme aggression, it was clear why the bottlenose dolphins had rakes, cuts, gashes, bruises, ragged dorsal fins and rostrums. Not only do they dig in holes and under reefs, they also beat each other up on occasion. The temperaments of the bottlenose and spotted dolphins were apparently quite different and they mediated their conflicts differently. Just a reminder that a dolphin is not a dolphin is not a dolphin and can be quite different across species, or even within species but in different environments: something to keep in mind when we think we know who "dolphins" are. The BBC used this footage to show an example of escalated aggression in their film, and I still have the images, and sounds, burnt into my mind and into my damaged eardrum.

1996—Crossing Boundaries

I was now finishing a decade of fieldwork and although I had hoped by this time to have cracked the code of dolphin communication, I had just begun my breakthroughs in understanding. My logbooks were full of notes on interspecies interaction, from the bottlenose and spotted dolphin relationships to the spontaneous interactions we ourselves were having with the dolphins. Some of these interactions led me to think about really building a two-way system of communication so the dolphins could better interact and communicate with us in the wild, should they wish to do so.

The more we worked with the dolphins the more we saw their individual personalities and styles emerge. Barb and Stubbet, our northern underbite gals, had a habit of sneaking up on people in the water, startling the person and probably causing endless enjoyment for the elfish dolphins. If Marc Bekoff is right in his book *The Emotional Lives of Animals,* many animals have a sense of humor. We had seen signs of jealousy when Barb occasionally tried to wrap her fluke around my leg and then pushed me gently from underneath. Stubbet had shown me an example of tail-guiding this way (a behavior mothers do with their calves). Trimy then dashed in to chase Barb away rapidly, jealous of the human attention.

We ran the now familiar edge of the sandbank at dusk to meet up with familiar dolphins while they slowly drifted into the deep water to feed. And we would sadly watch the misbehavior of small boats that ventured up to see the "friendly" dolphins and we tried to educate them in hopes of minimizing any interspecies unpleasantness. Sometimes we left the area ourselves to reduce the pressure from what was now becoming a too familiar sight, new boats. I counted twelve other boats anchored up at our little dolphin wreck one day. It looked more like a parking lot in the middle of the ocean than the wild.

Like other summers I awoke at 6:00 A.M. and sometimes noticed the dolphins stationed at the bow already. I stumbled into the water and watched the now familiar courtship behavior. Liney chased Little Ridgeway inverted and mated with her. Such behaviors were normal and easy to recognize.

It became clear that when males were teaching younger males they disliked interruption. When Mugsy placed herself in front of Punchy and Big Wave, who had harassed her son Mel, the males chased and bit her. The males repeatedly tried to engage Mel but Mugsy was in the way. Finally Mel went limp as the males buzzed him; what I now understood as a signal of submission. I had seen another example of this "passive" behavior when Everest, a young male, chased Rosebud in a courtship attempt. As he approached, Rosebud went limp and passive. It was a summer of patterns.

I had also watched Leo with interest during his development from a calf to a juvenile. A northern son of Lilly, Leo was recruited by a southern group of males. We had just finished a snorkel on a shipwreck when we sighted some frigate birds to the east with a large group of southern spotted dolphins of mixed-age classes. Leo was here but just a few days ago these southern males were up north, head-to-head against Leo. I wondered if males from one group were recruited at a very young age to a neighboring group. By the end of the summer Leo was with a group of older males in the southern group. Every summer we saw the influx of a few new juvenile males, but from where we didn't know. I suspected that this was one way of avoiding inbreeding. If, at this age, a male moved to a new territory, or joined a nonrelated group, then by the time he grew up he could safely mate with the females in his now familiar group. I had always thought it strange that we saw males mating with females in the same group even though much of it was probably just practice. It made sense that these males, although identified with their current group, came into the group at a tender age, making them nonrelated but eventually familiar.

It was hard to beat the clear and complex example of male interspecies interaction we had observed during "Stubby's revenge." But we continued to see other types of interspecies activity, which included a female bottlenose dolphin named Mine, who darted and chased away the advances of Navel, Zigzag, and Rivet. Just like my observation, in 1993, when Latitude took shelter with an older male coalition when his mother Luna chased him, I saw Mel, Mugsy's son, take "coalition cover" when his mom tried to discipline him. It was nice to start recognizing these "patterns of behavior" that were consistent with developing juvenile behavior, even though different individuals were involved.

Such intimate interspecies social interactions begged the question of self-awareness. How self-aware were these dolphins? How could they not be aware of their own self when they were capable of such complex and intimate interactions with another species? It seemed a silly question but we thought we might try a simple experiment. To test self-awareness, historically researchers have used mirrors. Humans, chimpanzees, gorillas, elephants, and dolphins in captivity can recognize themselves in the mirror and explore their bodies in a type of self-examination. Would dolphins do the same in the wild? I recruited Dr. Ken Marten to help with an experiment. We rigged a makeshift mirror from a Plexiglas window by covering it with reflective material. The first time we took the mirror in the water, the dolphins were distracted by a bar jack under the boat. Katy came by the mirror a few times and kept her distance while eyeing it curiously. The second time we took the mirror in, Rosemole and Katy eyed the mirror and contorted sideways. Rosemole whistled and the five dolphins (Rosemole, Rosepetal, Katy, Little Gash, and Diamond) tightened up and swam around the mirror. One swam inverted and sideways, the way they checked out a human swimmer, and then they briefly echolocated on this new object. At first we tried to spin in the water with the mirror to keep the mirror side to the dolphins. Later we realized the dolphins were trying to swim behind the mirror to explore what it was they couldn't see.

The next day we took the mirror in with the same group of dolphins, but they were traveling and just kept going. One thing was clear, the mirror took second priority to the dolphin's agenda, much like our normal interactions, the way it should be in the wild. Dolphins in captivity have less to do, which probably makes them more interested in objects like mirrors. Had the dolphins already seen their reflection in the dome ports of video cameras or perhaps another dolphin's reflection from the surface, since you can't see your own reflection because of the angle? Dolphins are possibly self-aware in other modalities such as sound and touch, both excellent senses. It would be worthwhile to consider how an alien species would demonstrate self-awareness without a mirror. It might not be through vision and reflection, our human species-specific way of checking our appearance. We might temper our experimental tests to better reflect the species' sensory system. After all, we used to expect animals to speak English instead of learning their own communication system.

Since 1985 my fieldwork had focused on a central area of the sandbank. To the north and to the south were other peripheral dolphin groups and we know now that they are technically one community, overlapping and interbreeding. The northern group is considerably smaller than the central group, about twenty-five individuals but with strong and well-documented multigenerational families—the Snowflake family, the Paint family—and a few other families of unknown lineage including Trimy and her offspring. A third-generation family in the northern group, Paint, Brush, and their offspring, the underbite family, dominate. Through the years Snowflake had many offspring, some of whom have produced grandchildren. The northern group is small and tight, often found together but with little interaction with the larger community. For a dolphin this isolation can be problematic. Being on the periphery and less networked means they have less protection and less experience with social matters. One of the hazards of being a female in a peripheral and smaller subpod may be less exposure to critical motherhood behaviors.

Not all dolphins are equally skilled in instructing young dolphins and the mischievous behavior of a babysitter or an inexperienced mother can cost a young calf its life. Trimy, a member of the northern group, was first identified as a juvenile dolphin so I am unsure of her lineage, but this year we excitedly awaited her first birth. Trimy didn't seem to have much experience babysitting. One day she presented us with her first offspring, Tango. With a large notch in his dorsal fin Tango was easily identified. One sweltering day in July, as we swam laps around the boat for exercise, Tango dashed in from a larger group of five dolphins that were at a distance from our boat. I got a brief look at Tango underwater and then I heard some whistles as Tango headed back out to rejoin his group. I often saw Tango by himself as a newborn. "Shark Bait," I always called him, "that calf is shark bait without his mom." Tango did disappear quickly, most likely the victim of a vigilant shark and a not-so vigilant first-time mother. Until this incident, most of the first-time mothers I knew were very successful, and diligent about keeping their newborns close at hand. But Trimy didn't seem to know what to do with her calf. Since there were relatively few calves for young Trimy to take care of as a babysitter it is likely that she simply didn't learn those critical behaviors, and it cost Tango his life.

Trimy's next offspring was born three years later. Tyler, a young female, nursed for more than five years, well more than the typical three-year time limit of dependency. By the age of three, when the mother typically gets pregnant again, older calves normally move into young juvenile groups forming their own social bonds and learning the community's social rules. In Tyler's case, Trimy didn't get pregnant again, so Tyler stayed on as her calf and missed out on some critical life lessons with other dolphins. I think the fact that Trimy belongs to the smaller, peripheral northern group was also a factor. Peripheral groups often have lower social status and less access to food and protection. So it may be that Trimy simply had less exposure and experience as a babysitter and did not learn how to enforce disciplinary boundaries for young calves. These females have high calf mortality due to a lack of experience and perhaps bad nutrition and a higher predation rate.

After a decade the seasonal continuity of my research allowed me a glimpse of the transitions of individual dolphins through their physical growth and social relationships. The southern group is also peripheral to the central pod but interesting in its own right. The southern group has a few more members than the northern group and their home range overlaps with the central group. But the southern group is notoriously harder to find even though their home range is technically closer to land. Rocky shores and very shallow sandbanks near the island make it harder for our research vessel to survey the area. So we are always relieved when we find this group and take advantage of the chance to observe them. One summer day we unexpectedly came upon our southern group of dolphins near a shipwreck, a popular dive site and a territorial line between the southern and central groups of dolphins. Many mothers and calves were there including Infinity and Ios, Venus and Violet, and Flying A with a new calf. A young male coalition including KP, Mohawk, and a third male spotted dolphin were also present. Suddenly a large group of older males (both southern and central) started fighting with head-to-head charges and associated squawks and screams. Everest, a young male, swam with the mothers and their offspring. Symmetry, a juvenile female named for the symmetrical notches out of each side of her fluke, swam with Stubby, a central male. White Spot, a southern male, herded everyone around and then went head-to-head with Venus on the bottom. This is often what happened when the southern and central groups met, but the reasons for this overlap in territory remained a mystery. Was it accidental because they shared foraging areas or did it provide an advantage when mating?

Immigration also occurred between the central and southern groups of spotted dolphins. One morning we left our anchorage and headed into port due to poor weather conditions. Our route from our central study area took us through our southern group's territory. On the way down to the island, while still in deep water, we picked up members of the southern group including Flying A with Flashlight and her calf Flare and also a male, Horseshoe. Years earlier Horseshoe and Flying A had emigrated from the central group and joined the southern group. We drifted into two hundred feet of water, a bit unusual for our daytime encounters, and watched them in their new group.

It's always a precious observation to see a newborn calf. As you might expect the mother-calf bond is the tightest in the dolphin community. Females are pregnant around one year and they give birth to a precocious dolphin every three or four years. The calf stays close or underneath the mother most of the time and rolls awkwardly to breathe at first, slapping its head on the surface rather than a smooth roll. Initially the mother will present herself for nursing by rolling over on her side. The calf wraps its tongue around the teat and the mother squirts milk into the calf's mouth. After a few months the calf is the one to change position to access the mother for milk. It's typical to see nursery groups together, with three or four mother-calf pairs. Males sometimes guard the periphery of these groups that can also contain very pregnant females. Mothers watch behind and to the side and watch their calves from a distance. As the calf grows he learns to pick up subtle movements and intentions by the mother as she breaks and comes over to discipline him. Calves test the limits of their mother's, or babysitter's, discipline by temporarily calming down but then resuming their out-of-control behavior soon after the attending elder begins to leave. They are truly young mammals.

There are many strategies for mating in the mammal world. Some include competition and displays between competing males. Others involve investment of time by the older males with mothers and their calves. And then there are sneaky males. We find this phenomenon in birds, sea lions, and other complex mammal societies. During an observation of serious courtship of Luna, primarily by Big Gash who seemed to be monopolizing Luna, I watched Flash, a young male, sneak in to copulate with Luna while Big Gash was on the side fighting the other males. But as usual things were more complicated than I first suspected. Flash and his juvenile friends were also chasing and trying to copulate with Latitude, Luna's young calf. Play and practice were often interwoven with serious behavior and deciphering the true activity became socially and behaviorally complicated.

Many old males we knew by now—Romeo, Big Gash—had been "fused" since 1985, making them at least, from my estimates, around fifty years old in 2008. After twenty-five years in the field, Romeo and Big Gash are still around and are likely the oldest dolphins in this community.

By now Little Gash, Mugsy, and Rosemole, at around nine or ten years of age, had all matured and had calves. Occasionally we have noted late summer births, but that is not the norm. During the first year of life, more than 25 percent of calves are lost, most likely due to natural health problems or predation from sharks. If the calf survives she spends most of her time with the mother, but is exposed to the mother's friends and other calves throughout this growth period. Over the first three years the calf suckles, but learns to catch fish, especially the easily gathered flounder, off the bottom. Spotted dolphins, it turns out, can be lactating and get pregnant at the same time. As is typical, the calves make friends with other youngsters and rapidly become juvenile dolphins. For years many of the adult females, like Luna and Gemini, regularly produced calves every three to four years. We don't know how long female spotted dolphins continue to reproduce. We don't know if they are senescent in their latter years, and go through menopause or, like orcas, have post-reproductive years solely as grandmothers—keepers of the knowledge such as elephants. To date we have never observed old females in the society who are not reproductively active, suggesting that when they are done reproducing they do not live much longer.

Since sharks attack young calves, especially noisy and inexperienced newborns, we were always anxious for the first sighting of the previous season's calves. Part of a dolphin's education is learning when to be quiet and when to be vocal. Being noisy can be a hazard and one that we thought Freedom, Flying A's calf, would never learn. Flying A and I grew up together since meeting in the 1980s. Named as a juvenile for the horizontal *A,* along her right flank, Flying A became a mature and successful mother with KP as her first offspring. Some calves had more of a distinct signature whistle than others. Some had a loud whistle, others had a raspy whistle, but most had a smooth frequency modulated whistle. Years later, when Flying A gave birth to Freedom, the calf's whistle was loud and raspy, landing her in trouble her first year of life and losing her a large chunk out of her fluke from a shark attack. Amazingly she still swam fast but apparently not fast enough since she managed to lose yet another part of her fluke later in the summer, probably due to a second shark attack. Freedom was notoriously loud and even I could identify her whistle as she approached outside my visual range. Her constant beacon of sound was probably what landed her in the mouth of a shark and it's amazing to me that she survived the second attack. A year later we saw Freedom and this time she was noticeably quiet. As the southern group, including Freedom and her mother Flying A, rode the bow we saw a few bubble trails from Horseshoe and Flying A. In the water Freedom was surprisingly silent. Perhaps she had another close call with a shark and finally learned to be quiet—finally learning the real meaning of freedom.

Hypervigilance is something we have seen in the dolphins occasionally and it probably has extreme evolutionary advantages. In early May, Havana came by with a new shark bite on the right side of his fluke. Havana swam by, frantically whistling while he observed a nurse shark on the bottom. This would not be the last time I observed a dolphin react to a harmless nurse shark. Havana's friend Mel (Mugsy's son) was nearby and Havana's whistling also alerted Mel. Although the dolphin's reaction seemed overkill, by the end of our encounter a hammerhead appeared and the dolphins chased him away, all the while looking pretty unconcerned. Young vervet monkeys learn gradually to discern predatory from nonpredatory bird shapes in the sky, and they hone their visual cues through exposure and experience. It's probably not so far off from the dolphin learning process about dangers in the ocean.

After a decade of work one of the things that became more and more obvious was our integration into this dolphin group. Most of the time it was positive, as in the demonstration of their signals like tail guiding and when they incorporated us into their group. Romeo, appropriately named for being a lover, would not only solicit a hug in the early years (which I still don't encourage, but it was amazing to witness), but also touch our face masks with his rostrum while he made little excited chirps. Sometimes the males treated human females like female dolphins and through the years I experienced Big Gash herding us away from the boat, or Stubby and his male group buzzing us ventrally upon approach as if going through the courtship ritual.

I am convinced that dolphins, when they "scan" us, recognize what they see under our bathing suits. Sometimes male dolphins focused their attention on male humans in the water and on occasion a male human swimmer aroused the serious interest of a hormone-challenged male dolphin. One day the dolphins were around all morning socializing. Rivet swam by with his penis out as he swam circles around me vocalizing excitedly with his male coalition, Liney and Flash, in tow. Then Flash, the up-and-coming young male in the group, hovered vertically in the water and scanned the human male, a fourteen-year-old son of a team member. Previously the older males had chased away some young male juvenile dolphins, but their conflict had not escalated. But Flash's male coalition suddenly synchronized their movements into a ritual of action. Focusing specifically on the human male, the dolphins charged our group with open mouths and as the intensity built, Flash charged us, postured with an open mouth, and jumped over us in the water as he tried to engage us. I wondered if they viewed this young human male as a threat, or did they have a fluctuating testosterone cycle from their previous encounter? Either way we retreated, as is appropriate for visitors in a strange land. Surely the dolphins were smart enough to recognize that we were different, not of their kind, yet their social behavior and use of their own communication signals suggested that, in some cases, they were treating us like conspecifics—other dolphins. When we enter their world we are subjected to their social rules, and learning the appropriate "rules of engagement" with another species is perhaps the biggest lesson of all.

A few years ago there was some disturbing footage shown on national TV. Filmmaker Lee Tepley and his wife, Lisa, were in Hawaii and had entered the water to film pilot whales. What was shown on TV was a pilot whale dragging Lisa, who luckily was a good free diver, down toward the bottom. The whale finally released her and people speculated on what the pilot whale was doing. Was he having fun with her? Had he tried to discipline her? Some shows were labeling this footage "when animals go bad." What the TV failed to show was that before the whale grabbed her Lisa had swum up to this group (who apparently was in a hunting mode) and stroked one of the large males positioned in the rear. By doing so I believe she entered the world of pilot whale signals and the unknown etiquette of a species.

While I recognized many of their patterns of dolphin behavior by this time the uniqueness of individual dolphins, and their behaviors, were also becoming clear. Year after year White Patches was visibly pregnant, but never appeared with a calf. Romeo had a propensity for human females, soliciting their attention. Stubbet and Barb routinely snuck up and displayed antics that were humorous. The sweetness of mothers like Rosemole and Hedley was obvious. Although I had expected, after a decade, to understand their communication signals, what I didn't expect was how diverse their personalities would be as individuals in the society.

Perhaps an untouchable subject twenty years ago, the scientific literature abounds with both qualitative, and quantitative, measures of personality in monkeys, horses, octopuses, and recently dolphins. Call it individual temperament if you like, but the results are the same. Some animals are bullies, some are sweet, some are bold, and some are shy. A social ecosystem incorporates the diversity of personalities and helps balance the political structure of the society, simultaneously providing various roles and jobs in the group. We see it in biological systems over and over again and it's called diversity. We are seeing it in social systems and we will likely see it in intelligent systems. It speaks of continuity: continuity in physical form, continuity in emotions, and continuity in intelligence.

By 1996 there were two things in my mind that were now clear: first I wanted to add some better technology to the project, especially to start recording the dolphin's high-frequency sounds, and second I was more and more convinced that we could really try two-way communication with this group of dolphins in the next few years. I began recording high-frequency sound when colleague Dr. Whitlow Au, from Hawaii, came out with his high-tech toys. Often called "Mr. Sonar," Whit and I had dialogued over the years about what we were seeing the dolphins do in the wild with their sound. This field season we had received a grant from the Office of Naval Research (ONR) to study the sounds that penetrated the sand while dolphins were scanning and digging in the bottom. Human-derived sonar was created after studying bats and dolphins, the sonar masters of the natural world, but there were still many things we didn't know about echolocation. In fact, Whit told me later that after seeing some of my video footage of the Bahamas dolphins, the Navy decided to reassess their sonar theory that had been based on stationary animals, and they began working with moving dolphins to better understand the dynamic of active signal use. Whit designed a hydrophone array that we buried under the sand and we hoped the dolphins would swim over it while scanning the bottom. The array's cables were attached to our boat with an underwater camera mounted on the bottom to record behavior. In between bouts of seasickness, Whit stoically huddled under the hot and sweaty outside cubbyhole we had created for his equipment and sampled and recorded sounds.

Whit had also created a mobile array to collect the beam-pattern of a signal from a free-swimming spotted dolphin. This was much easier than getting a bottlenose dolphin to swim over a buried array, as we had tried, since the spotted dolphin usually buzzed new objects or us with their echolocation as they approached. In early July we put the mobile array in with Venus and Symmetry, two southern dolphins. They explored it, buzzed it, and we got a few signals as a result—just doing what they do naturally. It was to be Whit's fate that every trip he was on was a rough one. Huge ocean swells often plagued his trips, forcing him to heave over the side in between collecting sounds. This year Hurricane Bertha chased us from our study area in July, an early but nevertheless threatening storm, forcing us to abandon our attempts with this equipment at sea and sending us scurrying back to land.

. . .

My work is often a dance between observation and interaction. Calves, by their very nature, are curious and friendly. The mother, the wiser of the two, is in charge of setting boundaries and disciplining youngsters. As observers in the water calves often get interested in us, rather than what mom is doing. As is our rule we try to respect what is going on in the water with the dolphin group. If a calf is excited and trying to play with us we back off and discourage the interaction. If the mother signals a clear comfort level, or intention to engage us, we use this time to build rapport with the dolphins. Awareness of larger social rules is especially critical during courtship activity when female dolphins get distracted and welcome human contact, instead of the male dolphin suitor paying them a visit. It takes extreme caution when interacting in the water and requires constant vigilance to detail. The daily life of a dolphin is routine in one sense and complex in another. In any complex communication the importance of social context cannot be underestimated. It matters who the players are, who is watching, what their history and relationship is, and what their personalities are like.

Luckily I had started using an underwater video with hydrophone from the beginning in 1985. After my graduate work I was well aware of the importance of capturing as many communication modalities as possible, and sound and behavior seemed obvious. My focus was to film social events in a great variety of circumstances with a great variety of individuals. The reality is that we film what the dolphins show us. The advantage to hand-holding a video camera in the water rather than recording underwater video remotely from the surface is that you can actually direct the camera at a focal point of activity while looking around to note the larger context. This has been invaluable for my work and in the recording of complex dolphin behavior. For example, I may be focusing my video on a mating trio, but I can look to my side and see a group of juvenile dolphins mimicking the adult behavior. Of course I would try to scan the juvenile behavior to document its simultaneous occurrence, but this gives you a feel for the three-dimensional world underwater. And it goes on and on in the decision-making process underwater. If group A is interacting with group B, might group C on the periphery affect group A's and B's behavior? And if so, it would be prudent to film the third group in case subtle movements affect the main groups (versus just film group A and B and later try to interpret what they were doing without documentation of group C). Documenting the sequence of behavior as opposed to short snippets is paramount to understanding the process of behavior and interaction. If you want to study the dynamic of communication you must record that dynamic, not sample it in a static time period, and that means recording sequences of behavior.

Also critical is the continued documentation of the social history of the individuals, their relationships and their past behavior, over years and years. I have always called this knowing the players, and nothing has been more critical than following known individuals, their mothers, siblings, coalitions, during interaction. I have been lucky to have a relatively small resident group of dolphins to work with, around one hundred in residence every year (over twenty-five years we have identified more than two hundred individuals, some of which are still around and reproducing). The average group size is around eight dolphins, which makes following the action underwater with a video a bit easier than a large group of seventy dolphins (which we see occasionally). Blessed to have found a species that develops increased spots with age, I have followed age classes and developmental periods to a degree unheard of in dolphin work in the wild. Now with a "library" of videotaped social events, it becomes possible to assess how dolphins act during their developmental phases. Many of these behaviors and sounds I described in some of my early scientific papers, such as aggressive head-to-head postures and the use of squawks, still stand today. So with all these senses we can observe (and some we can't) how do we "interpret" a communication signal to understand its function? One signal can have multiple functions and context is critical. For example, let's take an open-mouth display. It's not as simple as saying an open mouth is used in an aggressive scenario. One thing I have seen repeatedly in the wild over the years is open-mouth displays by individual dolphins, sometimes in the context of a visual display to another aggressive dolphin, sometimes while chasing a female, and sometimes by an isolated individual. Does this open mouth have the same meaning when there are no other dolphins present? We can clearly correlate open-mouth displays with aggression and loud vocalizations during fighting between individuals. Years ago I had wondered if, while chasing a female, they are "tasting" the water. Males often follow females in estrus and they follow enthusiastically. We always joke on the boat in the summer and try to figure out who is the "hot" female of the summer. The males monopolize and follow around certain females every season, which can usually be correlated with a young calf the next spring. Dolphins may go into estrus, and ovulate, one to four times per year. Every summer we can predict who is potentially in "estrus" not only by female behavior but also by the interested males. What is it that the males perceive? Is it chemical? Is it behavioral? Some mammals, including humans, hide ovulation, although not completely the pheromone scientists would tell you. Female chimpanzees display bright red genitalia as they reach maximum fertility. What do dolphins display? Is it physical or do the females get more tactile, less vocal? We have tried to think of ways of verifying estrus. In captivity researchers have measured estradiol and progesterone, emitted in female urine and feces, during these receptive cycles to correlate with behavioral displays. Are the males "tasting" her receptivity through chemicals from urine or feces? It is unknown if the male dolphins utilize these chemical cues like their terrestrial counterparts. "I would like to find a way," I explained to one of my masters students, Lillian Welsh, on the boat, "to drag a device, or test-strip, behind a 'hot' female this summer, or sample the water and see what might be in it that is measurable and potentially detectable to those male dolphins following her." But it is a challenge even getting close to a "hot" female with her male coalition of anxious fathers-to-be following close behind.

When there are no other dolphins around to display to, and the open mouth is not directed at a human in the water, what are they doing? One possibility is the dolphins are listening through their mouths. It sounds funny, but in fact dolphins are not only exquisite directional projectors of sound but good directional receivers of sound through the lower jaw. Recent work in this area suggests that dolphins may "hear" through their teeth and throat. This would be very handy if you were lost, trying to monitor the vocal behavior of a group at a distance, or monitoring for predators.

Most wild dolphin studies utilize surface observations for their behavioral work. Most marine mammals live in murky waters, a severe limitation for underwater observation. That is just the reality. It was clear that setting up a study in this area was a spectacular opportunity to observe dolphins primarily underwater and occasionally correlate it with surface behavior. When I am in the water I often follow the dolphins out of the water with my video camera, and if they leap, for example, I can at least correlate the two worlds. But it is underwater where they behave in detail and spend most of their time, and it is underwater where we need to look. Studying surface behavior only gives you a fraction of the whole picture. A researcher studying bottlenose dolphins in murky water concluded, by default, that the animals were foraging underwater and the sounds he had recorded were hunting sounds. When I heard his sounds I matched them with an underwater fight. Since he had no documentation indicating social behavior at the surface, and the dolphins were under the water for quite a while, by process of elimination he called these foraging sounds. Now I can tell you that it is often the case that dolphins fight underwater, for minutes at a time, and may never leap or come to the surface and show signs of their social behavior. Often they just return to the bottom, after a breath, to continue the fight. Certainly, other times they do repeated leaps out and back in. So, although it is complicated, I think the greater danger is labeling a sound with a behavior when you can't see it.

Decoding complex behavior involves context. One September day we had a group of nine spotted dolphins leaping and bow riding. When we entered the water Shorty and Ridgeway came in buzzing us intensely and then Shorty chased Little Gash with an *S* posture and then turned inverted and buzzed her. Now, if I hadn't known that these were males chasing a female, it would have looked like discipline with one dolphin upside down chasing another, such as a mother chasing a juvenile off-spring. I am a firm believer in describing behavior first as a naturalist and then measuring it. Without a sense of who the players are and their relationship, there can be no decoding of the function of such behaviors. I saw such inverted chasing behavior over and over again through the years: sometimes it would be a mother disciplining an unruly calf; other times it would be a male trying to mate with a female; and other times it would be a male trying to herd a stray individual on the periphery back into a traveling group. There was nothing more important than the baseline information on these individuals and the context in which they communicated to the interpretation of their behaviors.

Another keen example of the breakthrough in understanding their behavior came from observing examples of highly escalated adult behavior. Such behavior represents the fully actualized, normal behavior sequences of wild dolphins. Watching the development of these same behaviors in juvenile subgroups showed me the processes and contingencies of successful and normal behavioral development. Watching this complex system emerge over time, and across individuals, was the real prize.

Over the years I have come to realize that we think of interspecies interaction as odd, unique, different, and infrequent when really it is quite natural and happens every day between nonhuman and human animals. Fishermen in Mauritania cooperatively herd fish with dolphins, like their grandfathers did generations ago and also a generation or two of dolphins. The early Greeks and many other human civilizations have complex and interesting relationships and stories of dolphin encounters. Boys riding dolphins, dolphins helping sailors gain safe passage; the stories go on and on. With a complex brain already in place twenty-five million years ago, dolphins were likely the most intelligent brain on the planet at that time. After the evolution of our primate ancestors, human brain capacity surpassed, but only slightly, the size and complexity of the dolphin's. Did the dolphin brain evolve differently because of physical and social pressures in the water versus the same pressures on the land?

Like a scene from a *Star Trek* movie, we are heading to Mars, living on the international space station as we stretch our boundaries through the eyes of Hubble back to the big bang and the creation of the universe. We are finding the elements of potential evidence of life on other planets, including water and amino acids. More than four hundred exoplanets have been identified, and bacteria live in the bowels of the Earth where, until recently, we were unaware life thrived. What will happen one day when we stretch our boundaries and meet a perhaps sentient or presentient life-form elsewhere in the universe? Of course we have already encountered other intelligence on our Earth and our record is not good. Interspecies dialog, a true dialog, with another sentient life on this earth may show us a way.

At the end of my field season in 1996 I began thinking about starting my two-way work with the dolphins, Phase II. I felt like I knew their communication signals, the individuals, and after twelve years the dolphins still showed regular and spontaneous curiosity about humans in their world. The 1996 field season had not only seen the injection of new technology but had also brought a confidence and a reminder of the communication possibilities between our two species. I marked locations on the sandbank as potential spots to anchor the boat for Phase II that were shallow, somewhat protected, and where dolphins socialized; all elements that I knew would be important for two-way work. I had been thinking about interspecies work all of my life. Since I was twelve-years old I imagined what it would be like to delve into the mind of another species. What was it really like to be a dolphin? Who are they? How much is similar and how much is different? Is there a place to meet, to build a bridge across species differences? Could it happen in the wild?

I decided then and there to spend some time this fall and winter on land surveying my colleagues. Could I build a team to work in the wild?

What equipment would be involved? I already had a sense of how I would work with the dolphins. My training in graduate school and the well-known models of working with primate species already existed. I would build a system that was dolphin, and human, friendly. We would work in the water modeling, as humans, the communication system. If there is anything previous work had taught us, it was that a species does not automatically understand what you are requesting as a human. It's not how humans learn, and it's probably not how other social species learn. So we would use what other researchers, like Irene Pepperberg, in her successful work with Alex the African Grey Parrot, used—social rivalry. The idea was simple. Because social species like social attention you create an environment where a human models the system by acting "as if" they are that species. Or more simply put, everyone gets in the game, regardless of species. In this case, the humans know about the system and how to utilize it, and therefore act as models for the players. Little did I anticipate how well this would work with the dolphins.

PART 3

The Later Years

Insight—1997–2008

Two-Way Communication in the Wild: Is It Possible?

In 1997 I began pilot work on our two-way communication system with the dolphins. We had all the elements for developing human-dolphin communication. We knew the dolphins as individuals, we had observed their normal dolphin behavior for more than ten years, we had a catalog of their signature whistles, their fighting sounds, and most important, they seemed to be genuinely interested in us even after a decade. Although not every dolphin in the world wants to hang out with a human, this particular community of spotted dolphins seemed to have the interest and the time. Through the years we had seen examples of spontaneous mimicry of our behavior by the dolphins and we had exchanged other intimate moments. The dolphins solicited and incorporated us into their playtime, which involved keep-away games with seaweed or sea cucumbers. These "windows of opportunity" became our interactive time for our two-way communication work with the dolphins.

After four years it was clear that this communication was possible but our technology was still too slow, too encumbering, and too low-tech. The next few years I engaged colleagues, worked with voice recognition experts, engineers, and other scientists to improve our technology. The juvenile dolphins I worked with were rapidly becoming mature adults and were starting to focus on parenthood. I hoped we could continue our two-way work with the same group. But nature had other things in mind.

~ 7 ~

The Two-Way Work Begins: Phase II

> The subject of language has always been a contentious topic, scientifically but also emotionally. For both some scientists and laypersons, spoken language has long been held sacrosanct as being uniquely human, a defining character of what separates "us" (humans) from "them" (all other creatures).
>
> —IRENE PEPPERBERG, ALEX & ME

1997—Human-Dolphin Communication

In the summer of 1997 I began my work on two-way communication with the dolphins. We were going to give them a direct channel for sharing information, specific information about their world, and perhaps about ours. This phase had been twelve years in preparation. When I was a teenager I knew I wanted to study dolphin communication. I used to browse through the *Encyclopedia Britannica* and stop at the whale page. The blue whale is the largest living creature on earth, with a huge brain. What must they be thinking? Wouldn't it be neat to try and communicate with them someday? Little did I know that thirty years later I would be embarking on a two-way communication project with a group of wild dolphins that were now familiar.

More than a decade ago I came to this remote area in the Bahamas in hopes of finding a long-term research site to attempt interspecies communication with a wild pod of dolphins. I spent the first ten years getting to know the dolphins, learning to recognize their behavior, and giving them time to feel comfortable with us. It had been a wise investment. I knew their culture, their individual personalities, and their moods. Now I could attempt a long-term way of communicating with them with their permission and based on mutual interest. I was still as interested in exploring a nonhuman mind as I was when I was twelve. But now the reality was hitting. The last decade had been a hard road, building a base of relationship and information, but what if we really cracked the code, found a way in, and gave them a way to express their thoughts? What might this mean to our human species, the planet, to them as another intelligent species on the planet? To some it might seem insignificant. To others, such a feat was impossible, so why even try? To me it had been the burning question in my mind, one that science had only recently taken seriously, although still with an anthropocentric twist.

If I could ask a dolphin a question, what would it be? In some ways the dialog began for me back in 1985 when I first met the spotted dolphins underwater. What had developed over the years was a great respect and an understanding of dolphin life in the wild. But I still wanted to ask a question, make a more direct connection. What makes them happy? What do they enjoy? What do they think about? Do they have stories and myths in their culture? Some of these questions sound far-fetched but the more we understand some of the universals of communication or study the continuity and diversity of life-forms, the more it seems a reasonable thing to ask. Interspecies communication happens everyday between many species: between plants, nonhuman animals, and between nonhuman and human animals. What seemed to be lacking was our acceptance of our interdependence and mutual interactions with the natural world.

As I headed out to sea for another five months of fieldwork these questions were foremost in my mind. Will this year's first attempt at an initial system of communication be productive? Will the dolphins get bored, catch on, or be uninterested? Will we get frustrated? Will they? Will it take another ten years or will growing up with these dolphins give us insight unattainable any other way? As David Attenborough, the great naturalist renowned for his expansive nature explorations, expressed to me one night at dinner, after having spent four days at sea with him in 1990 on the BBC *The Trials of Life* series, "There is no replacing the experience and the time you have spent with the dolphins, the insights are ingrained, consciously and subconsciously. They provide a path for you in the future, sometimes without even knowing it." And although my strategic plan was to just spend time with them and see how they communicate with each other, little did I know how essential and critical a decade of baseline preparation would be toward this new step in communication. This summer I would find out.

During the winter of 1997 I decided to query some of my colleagues to test their reactions to the idea of a two-way communication system. Dolphin research, especially communication, is a bit plagued by old biases created around the work of John Lilly and many scientists have been concerned about being Lillyatized. John Lilly was a pioneering neuroscientist in the 1960s who decided to pursue an interest in communicating with dolphins. He reasoned that their big brains were used for something and knew they were both social and smart. Lilly pursued a controversial career trying to teach dolphins English and matching vocal output by the dolphins to human sounds. Lilly had designed a lab in Saint Thomas, in the United States Virgin Islands, where humans and captive dolphins could live together. Here trainer Margaret Howe tried to teach dolphins English. Although they were able to mimic the prosodic aspects of human speech, such as rhythm and intensity, the dolphins were unable to produce consonants involved in the production of English sounds. This was also the framework used in early chimpanzee studies, but like dolphins, chimpanzees do not have the appropriate anatomical arrangement to make the human equivalent of words. Lilly's experiment was interesting, but in the end his work was looked upon as a bit too alternative. Since that time the field has not been taken very seriously (even though work with primates and birds was in process albeit with its own challenges).

Even before Dr. John Lilly's work, Dr. Wayne Batteau began working in 1964 with the U.S. Navy, and attempted a two-way program of acoustic interaction with captive dolphins. Using up-to-date technology, Batteau cataloged existing dolphin sounds and arbitrarily assigned them to the human alphabet. More recently, the Kewalo Basin Marine Mammal Laboratory in Hawaii successfully used an artificial language to test the comprehension of commands from humans by dolphins, using a gestural and an acoustic system. A few semisuccessful two-way systems have been explored, including Dr. Diana Reiss's work with an underwater keyboard with captive dolphins, Dr. Ken Marten's underwater touch screen, and John Gory's two-way underwater keyboard at Epcot Center at Disney World in Orlando, Florida. These two-way communication projects are no longer active, but helped pave the way for further thought and development of the idea of creating a system for dolphins.

What has been clear in interspecies communication work from various research projects is that mimicry, although a potential mechanism for communication, does not guarantee comprehension of a language by another species. Anybody can mimic sounds or actions, but attaching meaning to words or actions comes only through repeated and meaningful social interaction and exploration. There have historically been accidental meetings between species, including the famous wolf boy, gazelle child, and others. In these cases actual interspecies adoption took place with the human adapting physically and socially to the other species.

The communication abilities between humans and other species have been successfully explored. Two-way work by Sue Savage-Rumbaugh with Kanzi, the pygmy (bonobo) chimpanzee, utilizing a rich and interactive keyboard and human companions is the best example. Through exposure to a keyboard that his mother was using, Kanzi spontaneously began to use the keyboard as a communication tool. Previous to the Kanzi work Savage-Rumbaugh had worked with common chimps, Sherman and Austin. One of the most remarkable, but usually unacknowledged, parts of the videotapes of Sherman and Austin doing tasks is the human researchers interjecting English words during the test such as "Sherman, go to the room and get the ball," and the chimps understanding them.

In fact the chimps have, by necessity and through functional interactions, learned to understand English.

Similarly, Irene Pepperberg's work with Alex, the African Grey parrot, successfully employed a framework called "social rivalry," essentially providing both a model and competition for attention during sessions and including rewards for using the correct words for requests and inquiries. Alex, too, could understand spoken English to a certain extent. And we need to remind ourselves that human children comprehend much more than they can express and articulate at a young age.

Three key elements are involved in these interspecies projects. One is that humans interact with each other and model the communication system with the nonhuman animal, rather than testing or demanding that a nonhuman suddenly understand what is expected. Second, using social interaction is preferable to using food rewards. Third, the use of an appropriate communication modality (for example, visual or acoustic) is essential to maintaining interspecies exchange.

I knew of only one attempt, in the mid-1980s with wild dolphins, which had shown promise for two-way work. A young student named Jody Solo, working with the late Ken Norris and his group with spinner dolphins in Hawaii, attempted to swim out to meet wild spinner dolphins using familiar greeting calls and nonaggressive behavior to gain the dolphins' trust. Although no extensive information came from this attempt, Norris believed that if Jody had an underwater video camera she might have returned with an intimate glimpse into the life of spinner dolphins in the wild. Norris attributed her acceptance into a wild school to Solo's ability to avoid threatening behavior and reach out with natural sensitivities to the dolphins.

I had already spent almost ten years in the wild with the Bahamas community of dolphins. I knew them individually; I knew their communication signals, their social relationships and histories. They were still curious about us. So when I decided to ask my colleagues what they thought of me trying two-way work in the wild with these dolphins, I was pleasantly surprised. With the exception of one colleague, most thought it would be interesting and possible, given the dolphin's friendly nature and my decade with this specific community of familiar individuals. The fact that we could observe their behavior underwater was an incredible opportunity, as I had already documented. When I visited Louis Herman and Adam Pack in Hawaii, leaders in the cognitive fields of dolphin research, they were so interested that Adam came out our first summer to help. Armed with solid knowledge of the dolphins, and encouragement from some peers, I was ready to begin.

Our Phase II trips were closed to the public. I invited and recruited some of our regular participants who were good in the water and able to work as a team. The first equipment I designed was a simple wristband that generated tones. I designed the tones with different rhythms, in low-frequency ranges, because these acoustic signals had to be both recognizable to humans and dolphins and within the ability of the dolphins to mimic. Humans are good at recognizing rhythm, but not at recognizing frequency contours. So I created a variety of tones that were short little "melodies" that humans could recognize. The wristband, which I could activate without looking, was perfect. I needed to be interacting and looking at the dolphins and the activity during Phase II work, not looking at a display or a keyboard.

Our strategy was simple. We, as a human team in the water, would use this system when the dolphins were around to model how the system worked, in hopes that they, as another species, would join in. In all the two-way work that had been tried, the most successful work with chimpanzees and grey parrots used this framework. The idea is that a social species pays attention to other conspecifics and interaction is more effective than food as a reward. Having an animal press a keyboard and get M&M's does not demonstrate how a communication system works, it merely equates that key X gets you candy B, and so forth. So I would use the two-way system with a human team in the water, demonstrate the system, and let the dolphins explore it, use it, and make mistakes with it, much like a human child would start to learn words in different contexts. Our biggest challenge was to design a system that was both human and dolphin friendly, complex but flexible, and to be able in real time and at sea to make changes that kept pace with the dolphins' exploration. But in 1997 I wanted to start simple and see just how the dolphins would respond and work with us and I knew we would learn from the dolphins themselves how to adjust our technology and behavior.

Our etiquette with the dolphins was also laid out. We would not disturb their natural behavior, but if there were interactive encounters we would engage the dolphins. If the dolphins were foraging, fighting, whatever, we would do our benign observations. If we entered the water and they initiated interest or a game with us we would move into Phase II work. Our human team was prepared for this plan and with a Start and Stop signal on my wristband I could alert the team. And if, during Phase II work, the dolphins began foraging, mating, or any other normal activities, we would end the two-way work out of respect for their own needs. Since I was in charge of the wristband and initiating signals I would make these decisions, based on my decade of knowledge about the dolphins' behavior.

June 10 marked the first Phase II trip, but it was off to a rocky start. I had my handpicked human team of long-term supporters and colleagues. The night before we left Florida to embark on this new adventure we had dinner at the Reef, our favorite local fish eatery in my hometown. We ate and drank wine until late that night and we were off in the Gulf Stream by midmorning. About halfway through our crossing to the Bahamas, Captain Will Engelby noticed something funny about one of the engines. By the time we reached West End to clear customs we realized we had to turn back to Florida to fix the problem. A twenty-day trip, the first ten days were spent on land with our human team practicing with Phase II equipment in the local Holiday Inn pool. In hindsight it was a good way to start. There were no strong currents or large sharks to worry about. We practiced our recognition of the tones on the wristband. Three tones in an up-and-down frequency and rhythm meant "initiate the keep-away game." Five tones meant "dive to the bottom" and so forth. After ten days in the pool we were all ready to head back out to the

Bahamas and try the real thing in the water with the dolphins. So on June 20 we crossed the Gulf Stream to begin our now shortened Phase II trip.

With my team practiced and ready, we began our work in the field. I will never forget the first encounter with the two-way equipment. We were in the water practicing with the Phase II device and two dolphins came by. I played the "start" and "dive" commands as we humans responded to the dive signal. White Patches oriented on the boat as if she thought the sounds were coming from there. She didn't seem startled at all, but instead curious, hardly even cautious, just contemplative. Baby, a juvenile female, swam inverted a few times as I showed her my wristband.

Ten minutes after we had tried these first acoustic signals, Baby and White Patches left and then came back with Tink, Caroh, Paisley, and two mother-calf bottlenose dolphins in tow. We initiated a keep-away game and used the system as the dolphins watched. Baby and White Patches had recruited these other dolphins within ten minutes of first exposure. Baby made excitement vocalizations as they swam away. It looked like the species reaction may be different; bottlenose dolphins seem to observe, spotted dolphins seemed to interact. Either way, it was good to show the dolphins our interactive system and how we used it.

Researchers have often speculated on whether dolphins, or other nonhuman animals, communicate details intentionally and displaced in time. This, after all, is one of the hallmarks of language: to speak of things not in your immediate space and make a plan. If so, then dolphins probably have some specific details of information, encoded acoustically, into their language.

Two days later Stubby, KP, and Priscilla came by the boat. Stubby was disciplining Priscilla, but as we started diving with the communication equipment Stubby watched. KP and Priscilla drifted out of sight. We started with the "go" signal; then I tried the "keep-away" signal with the scarf as the toy. The dolphins played intensely with us. A large group of male spotted dolphins came by and Stubby and his friends left with them. But after five minutes, Stubby and KP came back, loaded up with seaweed and sea grass; they had brought their own toys back! The dolphins paid attention to the sounds and even the box on my wrist as I kept activating the sounds. KP dropped his seaweed and tail-slapped intensely, then went to pout in the distance (which I describe as hovering listlessly in the water vertically and slightly arched), until he got attention from the other dolphins or us. Of course this was only the first step—exposing the dolphins to a system of communication that we used with each other—but at least it was something they could be involved in. An hour later we were still playing the game. KP was now the game master, getting jealous and pouting when Stubby got attention.

I was anxious to see if any of the dolphins had made an acoustic mimic of the sounds we were generating with our equipment. That night upon reviewing the underwater videotape and looking at the computer and dolphin sounds there were three signals that interested me. They came in packets and in rapid succession, and were in rhythmic parallel to our synthetic sound. Could they be mimicking the signal already? Is it possible? KP had vocalized excitedly, which we had heard in the water during disciplinary behavior, but it looked like KP was mimicking the "rhythm" of our artificial sounds, not the exact frequency signal. This, of course, was giving us insight into their use of the "prosodic" features of communication such as rhythm and intensity. Perhaps rhythm was as important to dolphin communication as were the frequency contours of their signature whistles. Or perhaps it was all they could mimic, or wanted to mimic. Either way, it was important. Synchrony and rhythm, both the physical synchrony of swimming and rhythmic posturing and vocalizing, are used during their own behavior. The way in which the dolphins responded and tried to mimic may be as important as a detailed decoded conversation. Did KP and Stubby immediately recognize our attempts at more sophisticated communication? I felt as excited as the engineers in the sci-fi movie *Contact* hearing that first, faint signal from the stars: "We are here."

The next day KP came to the boat with Dash, a young adult male, and five other juvenile dolphins that immediately initiated a game with a sea cucumber they had found on the bottom. In the water we found ourselves horribly distracted since Dash had a stainless steel fishing line wrapped around his tail, visibly cutting through, as he swam with a young group of dolphins. We had, over the years, observed males babysitting younger dolphins but it was rare. I suspected that Dash was a bit limited in participating in his normal male activities and so took another role in the group. We had heard, from other boats in the area, of a dolphin with fishing line around his tail and this was the second time we had seen it ourselves. We actually had a plan in mind and decided to focus on Dash, not KP, which was frustrating for Phase II work. But Dash's life was potentially at stake. The fishing line was cutting deeply into Dash's tail, which looked infected. If it cut through the tailstock completely he would lose his fluke or worse. Although normally we don't like to touch or habituate the dolphins any more than they already are habituated, we decided to try to get Dash comfortable by showing him the wire cutters in the water in case he would let us get to the wire and remove it. I didn't expect immediate access but perhaps over the summer he would allow us to help him. We tried swimming close and stretching out our hands. We tried to demonstrate the wire removal on a human. Meanwhile KP was getting frustrated trying to initiate a game and was waiting for our attention. We briefly played with the sea cucumber, but the two-way communication had to wait.

I can't help thinking that if our communication had been much more developed we might have been able to communicate our intention to help. There have been incidents in the wild with large whales where humans entered the water to remove a fishing line or other obstruction. The whales seem to sense the intention and remain passive, allowing their human neighbors to complete the task. So, perhaps it was not an impossible mission with young Dash, but today was not the day.

It was late June and we had spent three days anchored at the dolphin wreck. The skies remained clear and the water calm. Distant afternoon anvil thunderheads filled the sky, but we were untouched. Everyday the dolphins had been here, foraging enthusiastically on needlefish and ballyhoo at the surface. It was always impossible to observe them underwater during this behavior or initiate two-way communication work, since the dolphins were busy with their tasks. These last three days there had been no other boats around, leaving the dolphins free to play, search, chase, leap, and float in this vast expanse of the gin-clear water. We started our engines and began to head home to Florida. Three young dolphins darted over and stationed themselves on the bow ready for a bow ride. Soon all the dolphins, including mothers and calves, joined in. On calm days like this we are the wave generator. We gave the dolphins a bow ride for a few miles and we watched them break off from the boat to chase more fish under the nearby frigates, happy to see them enjoy their lives in the wild. These are the days we wait for, the calm ones—perhaps they also wait for these lazy days.

As we tried, once again, to begin our journey back to Florida, we ran into yet another group including Mugsy and her calf Martin, White Patches, and Baby. It seemed odd to find Mugsy without her friend Rosemole since these two females had been friends all their lives. But reproduction takes priority in a female dolphin's associations and since Rosemole had become pregnant, she hung out with other pregnant dolphins instead of Mugsy and her one-year-old calf. Female associations change dramatically by reproductive status, no matter how strong the previous friendships.

When we returned to the Bahamas we were anxious to try out our two-way technology again. We picked up anchor to find a group of ten bottlenose dolphins and twenty spotted dolphins. We focused our work on the spotted dolphins, including a large male group with some juveniles. Three dolphins dropped back to stay with us, including Paint and her male escorts Flash and Big Wave. Paint was pregnant and the males were protective and chased Paint while arching their bodies in an attempt to herd her. Paint approached me and did her vertical hang, the personalized greeting I always mimic. As Flash tried to chase Paint away she produced a bubble ring, as if she were annoyed with him, and came back to us to start a keep-away game. The trio of dolphins circled around us, observed, came in and out, and finally began to play with the scarf we had offered. Another window of opportunity had opened up for our two-way work.

It's July 4, 1997, and Pathfinder has just landed on Mars and the human race awaits the visual signals that should arrive in a few hours. The most interesting aspect of the Mars mission is the possibility of life on another planet. Are we alone? Is there someone out there on all those exoplanets that have been discovered? Sentient species on our own planet, those we already know of, can potentially guide us and show us the variety of intelligence that is possible in different environments. But are we ready to open up to other minds, beyond our species, on our own planet? It seemed striking that our two-way work should coincide with exploration beyond our own planet.

One of the most significant changes this year had been the collaboration with researcher Dr. Adam Pack. Coresearcher with Dr. Louis Herman at Kewalo Basin Marine Mammal Laboratory, and famous for their pioneering work on cognition, Adam agreed to join us during our pilot study this summer. Although I had met with him a few times in Hawaii at their lab and discussed ideas, neither of us knew if these ideas would hold sway in the field. "Experimentation" and "control" are not terms we use very often in our fieldwork, which usually entails benign observation. Adam is known for his cutting-edge work demonstrating that dolphins have cross-modal abilities between their vision and sound. Just as humans can visually recognize an apple and also recognize it by touch when blindfolded, dolphins can visually see an object and recognize it using their sound while visually blinded.

Over the years it had been apparent that we had rapport with the dolphins from their willingness to incorporate us into their games. We started recognizing some of the critical features of the interaction that molded our actions into a more formal structure regarding Phase II work. For example, we began to use eye contact and synchronize our behavior with the dolphins as a prelude to sessions. Once we felt in synch with the dolphins' actions they became maximally attentive to our actions and gestures.

We also began using pointing gestures to indicate interesting objects to the dolphins, such as a scarf. The gesture of pointing, critical in the development of human children and their abilities to make reference to objects, was a feature that Mark Xitco and colleagues at the Disney Epcot Center explored, utilizing divers to point at objects and locations for a two-way communication project with bottlenose dolphins. If dolphins point with their bodies, or by directing their echolocation beam, they may be able to direct the attention of other dolphins to things. This activity is considered a hallmark of sharing attention between individuals during complex interaction. By the end of our July research trip we had started to point not only to toys, but to encourage the dolphins to check in with us for objects. We also began to point to locations where interesting objects, such as sargassum and fish, were located in floating jars off the anchor line, creating a dolphin treasure hunt.

It was our initial hope to determine what interactive features were critical to get the dolphins' focused attention and also what objects or activities they might be interested in. Of course, the more interesting we could make the boat, the toys, and us, the more likely we would become playmates and have something to communicate about. Beyond all my expectations we had made clear progress toward the beginning of a two-way communication system. I saw how our human coordination in the water could be refined to further initiate contact and I could envision how dolphins and humans might have incentive to swim over to an underwater keyboard together and begin to explore the possibilities. I had been wary and unconvinced that an underwater keyboard would really be possible in the wild. Would wild dolphins touch such an apparatus? From his underground lab at Sea Life Park, Ken Marten had used a computer screen on the dry side of a glass window that the dolphins could trigger using their rostrum to break an infrared beam from their tank side. But the Bahamas dolphins were wild and they were not about to push a foreign object in the water. It might be possible to design an acoustically triggered keyboard that would activate when a dolphin explored or pointed to it, but that would be for the future.

By mid-July we were in the routine of setting up a "playground" around our anchored vessel. Toys were put inside floating plastic jars secured to our anchor line. One of the labels on my wristband was for "open," signifying a joint activity we hoped would encourage the dolphins to request a toy, since only a human could open the underwater jar and extract the toy of interest. While in the water playing my own signature whistle from the wristband (my own little whistle I greeted the dolphins with), Little Gash and Laguna came by with Havana. They were briefly chasing fish on the bottom, but Little Gash suddenly spotted the jar on the anchor line. I headed over to it and followed her as she oriented from the surface and then joined Laguna. They both dove very close to the jar and scanned it. When I looked back Havana was over my left shoulder watching me videotape Little Gash and Laguna. I then went over and tapped on the jar and tried to offer a demonstration.

Later that day Havana came by again with two other males, Mel and Everest. The dolphins dove down and solicited us so we started the keep-away game with tones and also a few dive signals. I saw Mel head over to the anchor line so I followed him. I hung by the jar, swam around it, and then when the rest of the humans and Havana and Everest came over, I signaled Adam (with two tones and a point) to open the jar. Adam opened the jar and got the scarf out while the dolphins hovered on the surface watching. So now the dolphins had seen us use our signals to communicate to each other how to open the jar, but would they pick up on this opportunity?

The next two weeks continued with calm, flat weather, a result of an El Niño summer in the Pacific that made for great weather in the Bahamas. Once again we entered the water and headed toward the jars on the anchor line while a large group of dolphins watched. We played the open signal and pulled out the scarf from the jar and as the humans played with the scarf Hedley and Heaven watched the activity. Then Havana, Heaven's brother, joined us as the whole Hedley family watched the human activity around the jar. Throughout the rest of the day the dolphins were in and out exploring the jars and toys around the boat.

Still later that same day the dolphins came by the boat for the eighth time. While Havana foraged and poked around in the sand, Dos dug in the sand and pulled out a snakefish while her calf Ditto attempted the same behavior. As I activated the dive down to bottom signal Ditto remained attentive even as an eight-foot tiger shark swam behind us. We kept working while Ditto made his way toward the anchor line with the jars full of treasure. After eight different encounters in one day it was clear that the toys in the jars had piqued the dolphins' interest.

Our initiation with new toys and focused attention seemed to encourage even more interaction with the dolphins. One morning Nassau and Caroh came right up to our anchored boat and hovered underneath the gate where we enter the water. As I got in first they huddled around me, bumping and rubbing me as they sometimes do. Caroh had a piece of sargassum on her pectoral fin so I handed off the video to my assistant and immediately chased them around for the dangling seaweed. Soon three of the dolphins had sargassum and I enthusiastically played the keep-away tone while we played. Suddenly one of the dolphins found a sea cucumber on the bottom, so we went down and dug up the sea cucumber and pushed it around on the bottom. Soon we and they were focused and competed for the new object of fun. Adam and I pointed back and forth, brought the sea cucumber up to the surface, and dropped it in the water column. The dolphins caught it, bounced it on the tops of their heads and bodies, and actually managed to keep it in the water column during some of the play.

There are some encounters you just can't forget and this was an example of an intuitive decision to just play the game and be spontaneous. It truly was an interspecies rugby match. Although it was forty feet to the bottom, both Adam and I were able to dive down and push the sea cucumber around the sand with the dolphins. Reacting in the moment, I decided to be a dolphin and I pushed my way into the group to get at the sea cucumber. It wasn't about touching the dolphins but about getting in the game, and so I just tried to think of myself as a dolphin in the middle of the cetacean roughhousing. At first the dolphins were not comfortable picking up the sea cucumber in their mouths, but after Adam and

I brought it up into the water column and started bouncing it around with our hands the dolphins started grabbing it with their mouths. It was a new game for them to keep the object in the water column, but for Adam and me it was a breathing necessity to bring the sea cucumber closer to the surface. Ironically, even without the Phase II technology, this was a great example of human-dolphin communication. It was the real thing, a simple game. Play creates pleasure, chemical pleasure, so it probably has an evolutionary purpose to it. But to explore a more detailed exchange between two such disparate species we would really need some advanced technology: we needed a human-dolphin translator.

We often left the jars on our anchor line all day, or set them up early in the morning in case the dolphins came by. We wanted to be ready to expose them to this treasure chest of activity. One July day Dos and Ditto came by. I could see Ditto heading toward the jars so I swam inverted toward a jar to get Ditto's attention. Then we approached one jar together. I dove down and opened it and pulled out the sargassum to show him. Dos was in the distance and she must have called Ditto over because a whistle got his attention. He headed toward his mom and they swam off together on the bottom. But Ditto definitely got a good look at the jar and the treasure I extracted for him and someday I hoped to see them mimic the open signal, to request humans to extract the toy inside.

The work was so exciting that sometimes we found ourselves a bit distracted and drifted too far from the boat. Strong currents ripped through the sandbanks and even for strong swimmers swimming up current with equipment was not humanly possible. As we swam past the anchor line with the dolphins I noticed that one of the jars we had left out overnight was on the bottom. I assumed it just had sunk to the bottom with water inside and we could fix it later. The current, which usually kept our boat pointed up current, was less strong than the wind so the boat was faced into the wind, but pointing down current. When we drift too far from the boat it is usually off the stern because of stronger winds than current. But today we were swept down current in front of the boat and were caught unaware as we worked with Tink, Paisley, and

Rosepetal, all major players in the two-way work. Visibility was also bad. The ocean whipped up frothy whitecaps as the deck team struggled to keep us in view. After the dolphins left we floated and chatted on the surface about our successful session and our safety predicament. Suddenly our husky first mate let out a high-pitched cry as he felt a large creature bump him. Our eyes bulged and adrenaline pumped as we swam up current back to the boat, anxious to reach the mother ship as we instinctually formed a tight unit resembling a dolphin pod. Safely back on the boat we examined the jar from the bottom I had noticed earlier. There was no mistaking the hole; there was a ten-inch jagged shark bite in the jar. Little did we know when we drifted away from our boat with the dolphins that day that there was probably a large shark lurking in the now murky tidal-driven water somewhere, having already sampled our tasty plastic jar. Once again we considered ourselves lucky, but were reminded of our constant need to stay alert as humans in the water.

Individual dolphin personalities can make or break work both in captivity and in the wild. Our star player for the two-way work was Caroh, a maturing young female. In late July, Caroh, Little Hali, and Uno came by and Caroh broke off and swam over to Adam. Caroh hung vertically with her rostrum out of the water (often a solicitation) and synchronized her swim with Adam. (This initial synchrony of dolphin and human was a critical indication of a successful session.) Adam pulled out a scarf, but Caroh already had a piece of sargassum on her right pectoral fin and a game of sargassum chase rapidly ensued. Adam pointed to the scarf and Caroh gently grabbed the scarf out of his hand as she did mine later. Lisa, our cook, entertained the dolphins with sargassum until it was time for us to move to the jar. I knew having a practiced and communicative human team was critical to the work and today it couldn't have been clearer. Adam swam over to the jar as we had agreed to do the next time the dolphins were here for a session. The whole summer had led up to this moment. This was the window of opportunity: Caroh was here, we had our team in the water, and she was clearly focused and attentive. As I led Caroh over to the jar where Adam awaited I brushed her rostrum with the sargassum. It's in these moments that I have to wonder whether this type of interaction isn't more about the relationship than the task. Many people who work with dolphins will describe this sense and it may be that in the dolphins' world relationship is primary. I physically directed Caroh with my body, keeping it pointed toward the anchor line and taking the swimming lead. Then I pointed to the anchor line with my hand and gazed back at her multiple times as we swam slowly in that direction. Such intimate and specific events between humans and dolphins are often difficult to record. We routinely had our video in the water and Will's job right now was to stay with the action and record it. Captain Will succeeded in swimming parallel to me while documenting the whole thing on video, not an easy feat without disturbing the interaction. As we reached the jar, I opened it slowly for Caroh, and she watched intensely as I pulled out the toy to begin another game of keep-away.

Our work with the two-way system continued through August with good sessions and the same interested dolphins that we would label the "candidates." On August 13, we swam out to a group of dolphins that came toward the boat. We clumped up as a human group and rubbed our elbows together in the water, which is analogous to a dolphin's pectoral fin rubs. Rosemole and White Patches stationed themselves in front of us, facing away, but I could see their eyes looking behind their bodies at the interacting humans (dolphins have eyes on the side of the head so they have some visual field behind them). As they swam back toward the boat and headed to a jar on the anchor line I started to point to the jar while looking back and forth at them. Both Rosemole and White Patches and Little Gash and Laguna watched me take the sargassum out of the jar. Then we started sargassum exchanges between humans and dolphins. We returned to the boat as Rosemole and White Patches escorted us in the strong current that was sweeping us, once again, far offshore.

It remains a question whether dolphins have labels or words for things in their environment. They do have signature whistles, potentially a label or "name," since whistles are individually unique. Encounters with sharks and other predators afford us the opportunity to see if they have labels for types of sharks, such as a tiger shark, or a bull shark, instead of using a general alarm call. Researchers Robert Seyfarth and Dorothy Cheney have shown that vervet monkeys refer to the specific type of predator, affording them to take the appropriate action, such as climb a tree instead of run into the bush, depending on whether the predator is a raptor or a large cat. Labeling or having specific references and words for dangers in the environment may have survival value. We might then expect the dolphins to have evolved such signals in the wild. It is hard to get samples of these signals since we are potentially at risk ourselves from shark attacks. It is not unusual for the dolphins to surround us when a shark is present, or even escort us back to the boat with great determination only to find out from the deck watch that a large shark had also followed us back to the boat.

I left this field season with a sense of wonderment. Beyond my wildest expectations the pilot study for Phase II had been successful: the dolphins were still interested, a few individuals were especially willing to work with us, and it was possible to work with a group of wild dolphins with the help of technology. Most of all it reaffirmed my commitment to the etiquette learned during years of working with the dolphins. In one working session Rosemole and White Patches, females I knew since 1985 and both currently pregnant, stayed with us and followed me to the jar, watched me pull out the toys, and then played with our chosen objects. We had never expected pregnant females to engage in the two-way work because of their other priorities. Rosemole hadn't been exposed to the system at all, although White Patches had seen us work with it very early in the season. Rosemole and I have known each other since her first shark bite and my first shark bump, her first and second offspring, and her changing relationships as a mother. Through all her changes she was still curious about us and found a few precious moments to share.

As the days went by and we practiced our work in the water, it was clear that these researchers could work, dive, and think in the water toward the same goal. I watched through late summer as Hurricanes Claudette and Danny swept by without harm. We were a focused team trained like astronauts to anticipate every subtle move of our coworkers. In the water we naturally took our roles and worked as a team, one member engaging the dolphins while the other focused on getting to the toys to begin a session. We had a good group of dolphins who were the right age and sex (young female dolphins) and had the time and the personalities to work with humans in the water.

As we left the sandbank that summer I felt sad at leaving this beautiful place with such incredible dolphins. I wasn't anxious to get back to land and deal with the day-to-day operations of life as a terrestrial social mammal. Most of all I experienced confusion, and a bit of awe and wonder, at how the time had flown and another field season was finished. Soon it would be May and we would be back on the boat to see if we could continue the two-way work with better technology. I knew then, more clearly than ever, that the dolphins would be mutual partners in this adventure, an adventure that could lead to who knows where. It was an opportunity to study another mind, to communicate with it directly, and to know another species, perhaps applying this insight down the road to understand other species or other life-forms of unknown nature. I thought about how my life had built up to this point from the first time I opened my encyclopedia to the cetacean page, thinking about how a mind in the water might evolve and watching the underwater world of Jacques Cousteau.

There were many reasons why this season was so successful. First of all the weather was working to our advantage and it was possible, and even pleasant, to anchor the boat for two weeks at a time. Second, I had recruited a talented team of colleagues. As I watched Hurricane Erica approach us from the east I began thinking about next summer and the potential for the two-way work. Perhaps this really could work!

1998—Caroh, the Scarf Thief

As we crossed the flat and beautiful ocean in early May of the following year, I found myself outside trying to get my stomach to adjust to the sea and its rolling motion. Our stores were full and spirits high, but everyone was so exhausted from the last six months of preparing the boat, equipment, and office that the energy level was low. Today lots of naps were in order. Captain Will had managed to keep up his tradition of hurting himself the first day of the season. This time he dropped the engine hatch on a finger. It wasn't broken but plenty black-and-blue—a rough start for an already exhausted captain. The wind was out of the southeast, a normal summer direction. We headed up toward our favorite shipwreck for a quick snorkel to rebaptize ourselves in the water, but there was a dive boat on the wreck so we continued up to the south bar to anchor for the night. We arrived on the bar just as some late afternoon thunderstorms from the west were moving in from Florida. As we were about to enjoy a great meal outside along with a rainbow-colored sunset, the squalls moved in, interspersed with horizontal and vertical lightning, and a group of spotted dolphins swam by, an eerie scene but a fitting welcome for a tired crew. None of us slept that night due to the slamming and creaking of the boat as storms blew through.

The next day the weather picked up. We found a group of bottlenose dolphins that were curious and after photographing dorsal fins from the surface we got in the water. It felt like we were in a washing machine with the tide pulling us one way and the surface swells hammering our bodies the other way. The dolphins surfed the large swells, but unlike the dolphins, I struggled against the force of the waves as my stomach churned like a tropical storm.

After another rough night I decided to head up north before the weather got any worse. It was a good choice. We found two mom-calf pairs, Stoplight and Siren and Lilly and Lava, who had survived from last year. It was a relief to see that they weren't part of the 25 percent mortality rate of first-year calves. Then Gemini and Galaxy rode our bow for a while. On our way back to the dolphin wreck a group of juveniles came leaping into the bow. It was Lucaya and Paisley and a few others and they were foraging intensely on the bottom, scoring big on razorfish and chasing away the opportunistic little tunny that followed and harassed them for the scraps. As we picked up the anchor and began to move more dolphins showed up. This time it was little Rosepetal, newly independent from her mom Rosemole, who was probably around somewhere with a new calf. Rosepetal had a two-inch remora that held on for dear life while Rosepetal turned and spiraled on the bow.

After leaving these dolphins we decided to head south again since the weather was picking up. The forecast was bleak with continued small craft warnings and fifteen to twenty knots out of the southwest. We arrived at the south bar to find a group of nine spotted dolphins, including Rosemole and her new calf continuing her fine tradition of showing up on the first day. Stoplight was also here, having traveled ten miles south from where we saw her that morning.

After another tough night we awoke to see five-foot waves and increasing winds. We downed our caffeine, spotted our first shark of the season, a large hammerhead, and headed south past Memory Rock to take shelter anchoring behind Sandy Cay, our home during rough weather. The wind was brisk and steady and expected to last until next Tuesday but at least we might get some sleep. Seven-foot swells in the Gulf Stream were predicted by the NOAA weather radio and that meant it would be ten-foot seas here. The water temperature was already eighty-two degrees, very warm for this time of year. Eighty-four degrees marks the necessary temperature for a good hurricane to form. Hiding behind Sandy Cay, we thought up creative recreational activities to keep us from eating all the ship's stores. I broke down and attempted to make a cobbler. Well, at least it was fruit.

I found myself really anxious to get on with Phase II work after the great success of last year and wanted to get on with identifying the spotted dolphins early in the season, but also wondered how much longer I could do fieldwork. I needed a change, and certainly Phase II was part of it, but it was hard to start the season so slow and with such low energy. Every year I went through these feelings at the beginning of the season. Perhaps it was leftover exhaustion from office problems, but I was trying really hard to leave all that behind. Tomorrow was another day.

It was only May 9, and we already had nasty, nasty weather. It was hard to do any work even on the computer. I went swimming to get some exercise; nothing like fighting five-foot waves to give you a workout. Our first mate got back out because of bad visibility. He said he was worried about sharks, but I was so wound up I stayed in the water to dissipate my energy with some physical activity. Shortly after I got out of the water someone spotted a large shark by the inflatable. The weather forecast was looking bleak and the seas were building in the Gulf Stream to not only uncomfortable but also dangerous conditions. Nevertheless we were productive, updating our dolphin identifications and calves and having a good shake down of the boat and equipment. May is always a transition month, both for weather and for my mental state, fluctuating from turbulent to calm.

In mid-May it was blowing hard out of the northeast at a steady twenty knots. The deep waters of the Gulf Stream emanated a vibrant grape purple, a deep dark blue violet that draws your eyes down to unfathomable depths. White frothing waves frolicked at the surface while seagulls dove into the water to secure their supper. Turbulent but magical, powerful yet peaceful: it was the synopsis of my life and I breathed easy taking it all in. It was far from calm, but instead dynamically alive. The air was fresh like the taste of a first snowfall. I was at home there, at that moment, and not even the dolphins could replace or compete with this experience. For it is a home of environment, not companionship. I thought of Jane Goodall hooting in the rain and I sighed with such completeness it was almost beyond my abilities to comprehend the feeling. I reveled at being in this element every day, every year, breathing, feeling, and hearing the ocean in its rhythms. It's like nowhere else. I am drawn to it because of my work, but it has become a destination in itself. Leaving land is a natural process for me, along with a desire to breathe fresh air and watch the dynamic ocean mold and form in real time. Were the dolphins not at the other end I would still relish it for its own feeling. I suddenly realized that I feel most at home when I am leaving port.

I do my best writing in the water, especially on days like today. It was mid-June and a high-pressure system had just moved in. The last few days had been ridden with midafternoon thunderheads. Today the water was aqua blue glass interspersed with greens, purples, and pinks—a perfect day. We saw dolphins all morning as we made our way to the dolphin wreck. They were on a mission, lazy, swimming slowly, but with a purposeful direction. We watched as the rambunctious youngsters came to the bow to play, most likely without their mother's full approval. Laguna, a new calf of the year, frolicked, leaped, twisted, and turned underwater as if to say, "Look I'm a dolphin, I know how to be a dolphin and I'm going to show everyone." Little Gash patiently swam over to Laguna to put her on notice for leaving, and with a brief touch and direction change, they swam away.

I never really thought of our etiquette with the dolphins as an agreement until I read Jack Turner's dark but honest book *The Abstract Wild*. He gives a multitude of examples of covenants, agreements, albeit nonverbal, with other nonhuman species—an arrangement of place and space. That's what we've been doing for the last decade. I called it etiquette, etiquette of trust stemming from respect for their lives. No feeding, no chasing, but "go on your way to feed, nurse your babies, it's okay, we'll be here when you are ready to interact and show us your lives." I've always looked at it as an investment in a relationship; respect and trust gets you in and harassment keeps you out. Human anthropologists know this, they observe, they watch, they wait, if lucky they get incorporated into the internal rituals, community secrets, and empowered places not available to everyone. I remember having the privilege of meeting Mark Plotkin, author of *The Rainbow Serpent*. Mark had spent many years with tribes in the Amazon working as an anthropologist. So at this gathering at a funder's house I asked him how he would approach the dolphins, given what he knows of approaching a human culture. He said simply, "Just like you have been doing."

After watching various groups of dolphins head off to the southwest on their mysterious mission, we anchored the boat at the dolphin wreck. It was the weekend and that meant no dive boats in the area. It was like the old days. We were in the midst of a sandy ridge, a place where the dolphins sometimes swam by. Time ran slowly and days were long in between dolphin encounters. I got in the water to do ten laps. I paced myself around the boat, sixty feet each way. The water was ideal for swimming, no current and crystal clear. The sunlight shone down twenty feet below and patterns on the sand rippled. Concentric rings of sunlight overlapped and glistened, cross-hatching the bottom. The ripples of light reminded me of a stone called larimar I bought in the Dominican Republic when studying the humpback on the Silver Banks in 1981. Larimar is unique to the island with a sky blue face interspersed with white sunlight ripples. In 1988, after I finished my master's degree in San Francisco, I commissioned an artist to make the rock into a pendant. She created three silver dolphins in the images of Rosemole, Little Gash, and Romeo. Their silver representations bore their real-life nicks and cuts, and the rock they surrounded was as placid as the calm water was today. I gazed into it as I swam. It altered my state of mind for a brief moment until an eyed flounder emerged from the galaxy of sand ripples and created a shadow. In and out of the sand the flounder disappeared, much like my fleeting thoughts.

Nature out here had its own rhythm. We all had plenty of work to do, computers to type on, data to enter, pumps to rebuild, dinner to cook. But the most productive energy came from getting into the rhythm; early mornings of work, we rested when needed, we ate when hungry, wrote when inspired, did dolphin research when they allowed. I envisioned the dolphins placidly swimming through the water, probably still heading to the southwest. They broke the surface, experiencing the sensual movement of the water at the air interface. As they glided underneath the surface they passed through thermoclines, pockets of hot and cold, a sensation not to be missed. I felt those same thermopockets as I did laps around the boat. I am happy to bathe in the warm water, but exhilarated and calmed by the occasional cool, as it keeps my brain cells from frying.

As I swam in this ideal moment, I thought about the Jack Tar, the old hotel at West End, which was finally uprooted. The "bite site," where

Kelly Rossbach spent so many of her days looking for bottlenose dolphins, and being eaten alive by mosquitoes and no-see-ums, was transformed into a clay river of sand pouring around the corners of the old jetty; man reshaping the very belly of the earth. We were happy for the locals and for the renewed and replenished activity in West End. But there was sadness at the loss of the old. And there was fear, fear of too much growth, too quick. I again flashed on Jane Goodall's description of the destruction at Gombe, a chimpanzee habitat now surrounded by human development. I recalled entering the marina when a small inflatable with two men returned after an adventure. "We swam with dolphins," they told their wives in the marina. Is this a sign of things to come? More people, more pressure. I looked at a small cove on the west side of the island where I had once experienced a truly sacred moment; it is now gone. It was 1994 and we had a Japanese film crew on board, NHK, the BBC equivalent in Japan. We were in port for a break and I went for a walk to the south end of the island, which is shaded with native pine trees. As I rounded the corner I saw some dolphins floating and cavorting in the shallow, protected cove. I waded into the shallow water without mask or fins to see them, amazed at such a discovery. Was this a resting or birthing place? We will never know. Now it is someone's front yard and the dolphins are gone. The pine trees have been bulldozed down. This once sacred cove has just had bypass surgery. But what is it that we just bypassed? Sometimes we don't even know until it is too late. I know, I was there; I experienced this place where dolphins lingered and rested, a quiet repose from the deep waters. But the cove is now gone and another creature on the planet has had to adapt to human expansion.

I thought of dolphins in captivity. I read *Visions of Caliban,* a book about the chimpanzee plights, of humans causing the deaths of chimpanzee families to secure the now-orphaned offspring for zoos and human entertainment. How can we be so perverse and justify these things? The story of cetacean captures is an exact parallel—supply and demand. You demand the entertainment, we supply the dolphins. But we won't tell you how they are captured, netted, harpooned, or how families are split. We don't explain how dolphins are marketed, collected, and transported through countries, allowing our industries to state that they are not from the wild, not recently but just last year. Hundreds of dolphins are still slaughtered routinely in places like Taiji, Japan. While trainers and dolphin sellers pick the ones to ship to the Caribbean, to Turkey, to Asian aquariums, their mothers' and siblings' throats are slit and the water fills with blood and screams. It must be beyond horrific for a young dolphin to end up in a swim-with-dolphin facility after living in the ocean. There are many dolphins currently in captivity that can't be released and who spend their time swimming in small circles, waiting to die. What if we could create a retirement home for dolphins, like chimpanzees and elephants, where they could interact with people and still have a life of dignity?

Like the human psyche can express itself in opposites, so stands the contrast here in West End. The Gulf Stream—blue purple swells, rolling, vibrant, rhythmic in its wildness—is fresher and calmer than the marina, which appears calm but in fact is turbulent and in transition, uprooted and twisted from its nature. You will never uproot the Gulf Stream. It knows itself and cares not for being subdued. The marina has been molded, folded, and told what and who it is. It is still seeking the natural rhythms of tides and wind, but having lost its roots is itself uprooted. But no, you will never conquer the Gulf Stream; it will remain intact.

The season was off to a great beginning as we sighted some dolphins just south of the south bar. It was the southern group that we didn't see at all last year and as we got in White Spot, a male adult spotted, started mouthing a piece of sea grass. He was with some mothers, calves, and juveniles and apparently he was feeling playful. He dragged a piece of sea grass on his dorsal fin while another dolphin went to the bow with a big clump of sargassum on his head to solicit those eager humans on deck. Tim and I picked up on the game and started soliciting sargassum keep-away, one of the dolphins' favorite games. I handed off the video camera to an assistant to work directly with White Spot, as is my prerogative for Phase II work. We pointed at the floating clumps and then I tried pointing at the sargassum while it was on his appendage. The first time I pointed to it on White Spot's pectoral fin, he wiggled it and dropped it. Then I pointed to it on the fluke of Symmetry, a female calf, and she also wiggled it and then dropped it. If only our fingers could echolocate, then it would be like a dolphin rostrum point. Venus had a new calf and Infinity also had a calf that looked at least one-year-old (we didn't see Infinity last year so maybe she was hibernating as a new mom). Some of the other southern calves were here, too, including Flare with Flashlight, and Roxy; a great two-and-a-half-hour encounter.

I had Captain Will give the dolphins a good bow ride while we humans remained in the water. It's a thank-you and they clearly relish this pleasure. If we aren't fast enough to give them a ride they start tail-slapping at the bow. And in this case it seems to mean, "Bow ride now, please." It's important to respond to these requests for Phase II work because we take their stationing at the bow as an intended request and give them the reinforcement of the ride itself. So we stayed in the water while the boat darted around giving them their ride. Here we were empowering the dolphins to train us to consistently respond to a request from them. Animals prefer to control their environment. In fact, environmental enrichment became an important way for zoos to maintain the mental and physical health of captive animals. Of course wild animals are used to being free and having choices. What if we could associate an acoustic symbol for bow ride that the dolphins could mimic and request at will? It would probably be more successful than a request for a sargassum game. I suddenly realized we were going to need a bigger fuel budget.

In early June we set out on our third trip, but the first full Phase II trip of 1998. I always scheduled a few early spring trips, before June, to catch up with the dolphins, the calves, the spotting, to update our identification catalog. But now we had many regular Phase II team members on board, including Chris Traughber, Diane Ross, and Suchi Psarakos, in addition to crew members Nicole Matlack and Will Engelby. The first few encounters were typical, with some social behavior and some travel. We spent the first few days testing our equipment including our new digital camera and our buddy phones for human-human communication underwater. We practiced our play, we practiced our eye gaze, and we practiced being dolphins.

We finished a great day with the dolphins and saw Apple with her new calf briefly and then anchored at the dolphin wreck and waited. We were here. We were ready. "Come when you can," I thought to the dolphins. As it got dark, Will, Chris, Nicole, and I got in to do laps around the boat. A barracuda was a bit too inquisitive and did laps with us and we joked about the possibility that he was interested in Phase II work since he was checking out our scarf during practice earlier. As we were swimming around a large gray shape darted in to the middle of us. Heartbeats fluttered until we realized it was a dolphin, and not just any dolphin—it was Caroh! We hastily got the scarf from the deck and began a game of keep-away with her. It was so natural that I didn't even think about the video. We pointed and she picked up the scarf and played keep-away with us. As we drifted toward the bow she continued to play with the scarf. On her way down to the bottom she briefly checked out a sponge. I decided to dive down to get it and in the process became intrigued with it. I bounced it around in the water column wondering how it would be as a dolphin toy. Chris dove down and checked it out with me and we started playing with it. The other *Homo sapiens* on the surface watched us and no one was paying attention to Caroh. I looked over and there she was, full body orienting and pointing to the scarf that was drifting on the bottom just in front of her. She didn't budge, didn't come over to check out the sponge, she was pointing, with the orientation of her body, to the scarf. Her eyes moved to look toward us, but her body was a torpedo focused on the scarf. I interpreted it as a point—an extended, emphasized point—and I went and picked it up and we all began playing again. That was the closest thing I had seen to a point from a dolphin. I would also speculate that it was a stubborn point since Caroh hadn't moved or budged. She wanted to play with the scarf and nothing else was acceptable. And she got her message across. Although it was quite spontaneous to play with the sponge, in hindsight it was a great test of her ability and her tools for communicating her preference of activity. It was the first interactive encounter we had had with Caroh since last summer and it felt like we had just picked up where we had left off, already making progress and adding a new dimension to what was possible between a human and a dolphin.

Now, if I were reading this from the outside I might wonder what was really so significant about this activity. Perhaps it is just amazing to me that when we really take the time to look, animals are often communicating a lot of information. Was it really different than my cat going over to her food bowl, looking at me, and then looking at the food bowl? Well, Caroh was a wild animal and she made a personal choice with two objects that her human companions were playing with. She also swam away with the scarf at first (we thought we had our first human toy casualty) but actually she came back with a younger dolphin and reengaged Chris. Most amazing is that it was the first encounter with this individual, Caroh, who was our star player last summer. The pace was relaxed and natural. These subtle interactions were significant keys to the process of interspecies interaction. It had everything to do with synchrony, entrainment, pace, detachment and alertness, and nonattachment to the outcome, while being alert to the many possibilities and direction of events. These "windows of opportunity" became even more important throughout the summer.

By mid-June the days got so busy between watches, dolphins, and physical exhaustion that it was hard to write up notes. We had more encounters with the same group of juveniles including Caroh. We lost a scarf as Caroh swam away with it, possibly in the context of competition between Tink and Caroh. We also introduced the rope as a toy and Caroh loved it, spiraling it around her body in a sensuous dance. So far Caroh was the only dolphin to swim away with the toys. This morning we had another incredible game with pointing—exchange, rope play, and Caroh swam away with the rope. Interestingly, Rosepetal came back and was scanning the bottom for something, either fish or perhaps the rope, and then she left and shortly came back with the rope, dropped it, and left.

Did Rosepetal really have the intention of returning our toy to us? Had Rosepetal recognized this rope toy as a human object and returned it to us, unlike Caroh who continued to swim away with our scarves? We joked that Caroh was starting a scarf boutique somewhere in the Bahamas and we retired at midnight and dreamed of red scarves floating together in a giant sea quilt.

It was clear that the dolphins were attending to and watching our points at a distance; that had been firmly established with Caroh and a few other individuals. Luckily they had been coming back to the boat to interact all week, but it was moving fast. It was clear that we should just plant ourselves here, even in rough weather, because getting repeated exposures and time with the same dolphins was critical. With attentive groups of juveniles we began to introduce visual symbols that we would eventually mount on our human-sized keyboard. I had designed simple shapes of white plastic, about one foot in size, including a cross, a half-moon, and a star from hard white plastic. These were the shapes we began to associate with various toys. Rosepetal was emerging as a potential prime candidate and we decided to start paying more attention to her. Caroh remained the most attentive dolphin now, but she was maturing into her potential motherhood years with new responsibilities. It was incredibly lucky that juvenile groups had a range of ages, from four to eight years, because younger animals like Rosepetal could be exposed to the system for three or four years. Their own social learning also occurred at this age and their exposure to social skills and rules was natural.

We discussed at night, after watching old videotape, how many of the captive studies had attempted two-way work, but ended up with little or no results. Concern over such an attempt in the wild dwelled on the uncontrollable nature of free-ranging animals, so why not try it in captivity first? But perhaps the wild had some advantages; wild dolphins were more likely to be "fully actuated dolphins" with social networks and built-in behavioral complexity. Could it be even more likely to have success with dolphins in the wild because of their wildness? Our most critical issue in the Bahamas was to get enough time with the same individuals over and over again, but that was happening. We had been lucky with good weather the last two summers, which allowed us to stay anchored in one place where the dolphins could, and did, find us. It was more important to understand how the dolphins wanted to communicate in their world than it was to understand how we, as humans, thought they should communicate.

By now Caroh had been officially labeled our "scarf thief." It seems she had developed a taste for swimming away with scarves. The first two weeks of June were infamous for her scarf-stealing behavior. On June 10 we were working with three dolphins, including Caroh, on the two-way system. We started with the red scarf and had some good exchanges and then we tried a switch to sargassum play. Caroh was the main player with the scarf. We did a switch to sargassum and engaged Priscilla and Mitsu while Caroh swam off and came back and vertical hung at the surface with the scarf, before she joined our sargassum game. I tucked the scarf in my bathing suit and tried to play with the sargassum instead. When I dove down Caroh followed me and pointed to the scarf in my bathing suit, trying to get at it. Diane and I did a synchronized dive with the rope and scarf, and Caroh followed us down toward the bottom. Then we dropped both objects and pointed to the scarf, but Caroh chose the rope and later swam away with the scarf again. Thief!

Some of the technology we used for Phase II facilitated our abilities to hear the dolphins in the ultrasonic range as we communicated between our fellow humans. Our ears were made by Ocean Technology Systems (OTS) and consisted of earpieces and a throat microphone, allowing me to broadcast simple commands to my human team members in the water to coordinate our activity. Initially designed for a scuba system, I had modified my ear to have throat-linked, snorkel sound transmission ability. So instead of speaking real words in a mouthpiece for scuba I put the device on my throat and spoke into my snorkel. This allowed me to code simple commands such as start and stop to humans in the water, much like Morse code. I used numbers of syllables and rhythm to impart extra information and clarity to separate commands. It also turned out that with these ears we could hear many dolphin sounds normally hidden from us, since they received and transmitted up to thirty-two kilohertz, twice our normal hearing. So clicks and buzzes became louder and more complicated with our ears on. They were a great tool for both the two-way human coordination and passive observations of dolphins. In fact putting two ears on in the water gave us an extra ability to locate the direction of the sounds underwater, something not easily done with our normal human ears.

By mid-June we had lost more scarves to the thief master and her friends. One fine day we saw a large group of dolphins ahead as we were approaching the dolphin wreck. After they all rode the bow briefly, Priscilla and Caroh stuck around and were immediately attentive, surrounding us in the water. They seemed to be looking for a scarf toy on the human swimmers as they scanned and poked at our bodies, but we were playing hard to get (as we were running out of scarves). Suchi finally pulled out a scarf and played keep-away with them. Diane did some pointing to the scarf on the other dolphin while one attended and gazed back and forth and finally dove down to grab the scarf. We then brought the visual symbol in the water, but we were too late and the scarf was mysteriously gone again.

There had been more boats up here this summer than last. Last year had been nice because many of the dive boats started working in Bimini, where the dolphins are a bit closer to land, so it relieved the pressure up here, although maybe not so for the Bimini dolphins. We remained focused on staying anchored in one spot to ensure that the dolphins knew where we were for our two-way work and luckily the weather had been cooperative.

The rest of June we added to our lost scarf tally and it seemed like it would never end. We wondered where this mysterious boutique was set up and how we might reacquire our scarves. Even with her friends Mitsu and Tink around, Caroh was still the head instigator of scarf thievery. One day Mitsu came by and Nicole headed to the anchor line with the scarf, where the visual symbol hung by itself, not yet on the keyboard. Mitsu oriented and followed her to the anchor line. Nicole pointed back and forth as Mitsu attended and then she gave Mitsu the scarf. After Nicole retrieved the scarf, Caroh came in and oriented on her favorite object. I pointed to the symbol and Nicole gave the scarf to me, but when I saw that Caroh was still watching I pointed to the symbol, back to the scarf, and as Caroh looked back and forth I gave her the scarf. Nicole followed her as Caroh once again swam away, having pilfered another scarf.

Dolphins sometimes initiated the two-way work only to be distracted by a plump sea cucumber on the bottom, one of their new favorite toys. However one day something amazing happened. During a two-way session Caroh and Mitsu left and came back with two juvenile bottlenose dolphins. Adam offered Caroh the scarf and amazingly not only did Mitsu play with Caroh, but the two bottlenose recruits also grabbed the scarf as if already familiar with the game. Eventually these dolphins drifted away, but it left me with a strong reminder that we must be prepared for anything during interspecies interactions. Planning for all contingencies during the communication with another species may not always be possible, but we should strive toward anticipating the unexpected. I had never observed play with an object between these two species. Could it be that Caroh brought the bottlenose dolphins with her to expose them to the system? Were they equal playmates in the dolphin world? We had always joked about the bottlenose dolphins being the observant ones, the cautious ones. Were they the best candidates for Phase II? I hadn't really designed our sounds and signals with the bottlenose dolphins in mind because they had historically been shy around us, but perhaps I would need to rethink things.

I thought about all the unique events that were occurring out here in the wild. First there was the active recruitment between the two dolphin species supporting the intimate and likely individual relationship these two species have with each other, and also the cultural transmission of information, which would likely take the form of sharing signals and passing down the information to possibly nonexposed individuals. The possibility occurred to me that the dolphins might begin bringing their own toys, such as sea cucumbers or sargassum, to the keyboard for labeling. Would they share them with us? We needed to be ready with our technology to label and incorporate new words in real time for more powerful interactions.

It was early July and the last two days had been a bit slow. The dolphins had been coming by, but they had mostly been quick swim-bys. This morning Caroh came over with Scqew, an older juvenile male dolphin, who was acting like a babysitter, and foraging on the bottom while Caroh and Paisley darted around us. There were other dolphins in the distance and it felt and looked like Caroh and Paisley had come in just to play, bringing Scqew with them. In fine tradition Caroh left with another scarf, the count now three blue scarves, one red scarf, and a blue rope. We were still working with pointing to single objects and hadn't put the newly designed full keyboard in the water yet, but we needed to soon. I predicted that as soon as a dolphin, possibly Caroh, learned the power of even one of the keys, she would bring other dolphins over to observe. Maybe that's what she was doing with those two juvenile bottlenose dolphins? Caroh had already demonstrated that she would bring other dolphins back to the boat, either with a scarf herself to show them how to play with humans or without a scarf to show them how to solicit such an object from us. Either way, there was apparently value in sharing this information with other dolphins and in the case of the other day with another species of dolphin. Was there a social advantage in the sharing of information? This feature may prove to be the most important emergent quality of the process of interspecies communication.

Caroh continued to engage us with the scarf and she often came right up to us and took the scarf from our hands. At one point, Caroh came up and I offered her the scarf and she took it right onto her rostrum. I chased her a while as the scarf fluttered around on her pectoral fin, but she ended up swimming away (with the scarf again). The surf was rough and the dolphins did not come back; neither did the scarf.

The idea of the keyboard was simple. Pair an acoustic sound with a visual symbol and the object itself. After all, the human alphabet is merely a bunch of symbols strung together to form words. We knew from

Adam and other cognitive researchers that dolphins have the ability to understand abstract concepts and use symbols. Although my original vision of the two-way system involved a keyboard on the side of the boat attached to an onboard computer, it quickly became apparent that, unlike in captivity where dolphins have less to do with their time, these wild dolphins were not going to hang around the boat and work with a keyboard. We quickly designed a portable keyboard that consisted of a long strip of aluminum where the three visual keys (a cross, half-moon, and star) were mounted on a waterproof box. Inside each of the three boxes was a mini-CD player with one artificially designed sound to match each specific visual symbol. The apparatus was triggered by the operator of the keyboard when either a human or dolphin pointed to the symbol. I assigned roles in the water. One person pushed the apparatus and operated the acoustic trigger when the dolphins or humans investigated the keyboard. One person took pictures, another video. Two of us, usually Adam and myself, were the main humans to interact with the dolphins and we focused on leading them to the keyboard and choosing when and which signals to use. After each water session I wrote up a data sheet describing who was there, the main interacting dolphins, and what sounds and symbols were used in the water. We often discussed what worked and what didn't after each session, sometimes videotaping our thoughts for future reference. What quickly became clear was that slow, synchronized swims usually involving eye contact and slow movement were necessary to begin a session. If we entered the water with the equipment and the dolphins were doing natural behavior we did not engage them. If, after their normal activities, a few individuals remained with us and started to interact, we engaged the system. Our commitment to not interfere with their normal behavior, even to engage in two-way work, was crucial and continued to be a guiding factor in working with the dolphins.

Initially we had designed the signals to represent objects we already played with like the scarf and rope. I added a key for sargassum and a symbol for a bow ride, hoping that they would be empowered to request a bow ride by using the symbol for this activity. The fact is that the dolphins were already adept at tail-slapping at the bow and lining up for a bow ride. They already had us trained, but that was okay with me. All I wanted to do is add a new option to the bow ride, an acoustic one that they could produce when they were not physically at the bow, but wanted to request a ride.

Our strategy in the water was to not demand that the dolphins do anything specific but instead give them time to explore the equipment and watch us using the same equipment and observe the results of humans communicating with the keyboard. This was meant to be an interspecies keyboard usable by two species. We would model this system, as you would naturally model human language to a child, just by using words in different contexts. The child is also able to observe the contingencies and controls around such communication. If a human in the water pointed to the scarf symbol they got a scarf. If they pointed to a symbol of an object that was not available they got nothing. Even mistakes made during our human practice with this equipment might give us insight into how communication could be encouraged and modified.

In early July we decided to take the apparatus in and try it using human demonstrations. Baby and Brush were immediately attentive and initially I led them over to the keys and demonstrated the rope symbol. What was very clear last year and again on this morning was that the dolphins became both vocally quiet and slowed down their movements when attentive. There was a large group of adult male spotted dolphins hanging on the periphery also watching, but the interaction was left to the juveniles. It was interesting to think about scenarios that might occur between the dolphins themselves; did the juveniles just meander over this way and the adults tolerate their exploration at the boat? Did the juveniles communicate their excitement or desire to come to the boat? Did they exchange specific information that something interesting was going on at the boat regarding objects and toys? It was hard to know, but it seemed clear that it was at least temporarily okay for the juveniles to venture over here while the adults observed and watched.

Another thing that seemed obvious was that the jars were really getting their attention. Day after day Adam or I led some of the dolphins to the jars and demonstrated the treasures within. One day it was Paisley and Little Hali. The next day Priscilla and Brush watched us open the jars. Sometimes these sessions led to keyboard work, other days the dolphins lazily drifted away. But they took the time to explore this new system with us. Sometimes their friends brought back their own toys, such as sargassum or sea grass, thus complicating our already busy session with a human-made toy. Who should we spend time with? Was it fair not to play with the dolphins that brought back nature's toys and only pay attention to the dolphins with the human toys? That was never the intention, but it did make focusing on a goal difficult in the water. Our plan was to get the keyboard apparatus over to the jars so that the symbols were more accessible to our work at the jars. Clearly we needed to build in more words to accommodate their already growing potential vocabulary of toys.

Even though we got in the water with some specific goals, ideas such as "let's demonstrate the system or use the visual symbols," we still responded flexibly if they changed the rules. It was tough for some participants to understand the difference between specifically training the dolphins as opposed to exposing a species to a communication system. Simply put, it's about giving the animals control and time to explore the consequences of communication. It would be akin to ordering a child to pick up the right glass that contained milk and then not giving it to them if they picked up the wrong one versus letting the child choose and discover whether that glass would be filled with milk or not. For the dolphins and like human parents, we modeled the communication system in real life. As humans in the water we did what the dolphins could do if they wanted a toy; we went to the keyboard and pointed and requested it and some other player, in this case human, gave us what we requested. In this system both species were welcome to join in.

In any society, human or other, there are individuals with distinct personalities and preferences. As the summer went on, major players emerged, including Caroh, Tink, and Rosepetal. However, on the days when Caroh was not present to take the lead as head ambassador, Tink filled in. Sometimes Tink brought her other friends Mitsu or Storm to watch and participate. The continuity of encounters and repeated visits everyday was a clear measure of their level of interest. Perhaps fifty years from now other humans will venture into relationships and interactions with unknown species utilizing what we have learned about the process interacting with these dolphins.

Although El Niño years bring us typically good weather and lack of hurricane activity in the Bahamas, this year we saw additional effects of climate change; Florida was burning up. Even anchored offshore sixty nautical miles from Florida the smoke had reduced our visibility to a quarter mile. I canceled our afternoon watch because it was too hazardous to have people breathing the smoke outside. This is one of the most pristine places in the Atlantic and we were sheltering inside the boat because of the air quality. I tried to imagine what it was like in Florida; it must have been an extreme health hazard. The dolphins, too, being mammals, must breathe this same air. I went outside occasionally to see if the smoke was gone and instead heard the sound of a dolphin's labored breathing. Luckily, I could go back inside, but the dolphins couldn't. Dolphins in the Sarasota Bay area on the west coast of Florida, where industry compromises air quality, have acquired black lung disease, evidenced through necropsies, and have become the coal miners of the sea.

By the middle of July we had demonstrated the keys and the communication symbols to the dolphins quite a few times. Of course Caroh remained a major player, and was now up to five scarves and one rope theft. It was interesting that on one of our last days, July 4, she took our fifth scarf and swam away with it. Later in the afternoon Adam hopped in the water and when he went under the boat there was a scarf mysteriously rolling along the sandy bottom awaiting retrieval. Later that day Caroh came by with Geo for our last encounter with Caroh and Phase II for the season.

Some weeks were very slow on the sandbank as far as normal dolphin activity. Perhaps the dolphins went somewhere we have yet to discover.

Perhaps there was a fish run nearby and being somewhat opportunistic feeders they decided to gorge themselves. Perhaps there was a large, hungry tiger shark in the area. During these periods we entertained ourselves with data entry and sleeping. This week we spent part of a day at the Dry Bar snorkeling on the reef and then headed back up to the ridge and had an incredible four-hour encounter with spotted dolphins and bottlenose dolphins. A tropical wave was moving through the Bahamas and was headed for Florida packing thirty-knot winds and nine-foot seas, so we tucked our tails and ran for shelter behind Sandy Cay once again.

Working with the dolphins was often on their schedule rather than ours. We waited, we prepared, we hoped, and then nothing happened for days. One day we listed the mothers and calves of 1997 and 1998 on our outside board to remember what information we still needed to complete our field season. We needed to sex Rosemole's new calf and Infinity's new calf. We listed Luna, Gemini, and others we needed updated photographs of or information about. Suddenly we found a group of twenty dolphins and it was the entire list less two dolphins. Amazing and a bit eerie. We sexed the calves of Rosemole and Infinity and rescored the tally board outside.

Individual dolphins had different levels of exposure and comfort levels to different objects during the two-way work. Late in July we had a Phase II session with some northern juvenile females, Snow, Brush, and her sister Pigment, and Pimento. After they played with one another for a while and received three bow rides, they brought sargassum over to us and engaged us in their game. They were quite cautious with the scarf when it appeared, even after we demonstrated its uses, but it was a good example of the difference in individual comfort levels and exposure. Brush followed me down in the water column to investigate the mysterious object, but she avoided touching it.

It was important to have continued contact and to track the individual dolphins' exposure and their reaction to objects for future work. Even if

Caroh, our star player, decided to physically push a key it might not be preferable for the majority of interested dolphins, given their reasonable caution to new physical objects. The dolphins on the periphery were often the most observant of Phase II players, either out of choice or by lack of involvement with other dolphins. Learning through observation may be just as effective as participatory learning for dolphins.

Although I had agreed it was important to experiment during the two-way work, I thought that the best strategy was to expose the system to the dolphins and watch for indications of what the dolphins were interested in. Roger Fouts, in his amazing book *Next of Kin,* described his journey in ape language work and the unfolding of his understanding by observation versus strict experimentation. Fouts set up a creative and social context for his work that incorporated the individual chimpanzee's personality, the importance of observation, and the ability to keep learning fun.

Science, to be productive, must be creative and collaborative. The synergy of minds working together is perhaps the most amazing experience any scientist can have. One day we set up a "trial" with Caroh where two scarves were held up and I pointed to the right one while I had Caroh's attention. She went to the left scarf and grabbed it. Now there were multiple ways of interpreting this experiment. First, it could be that Caroh was just not attending to the point and chose which scarf she wanted. Second, perhaps the other scarf without my point nearby was easier to access so she took it. Third, maybe to Caroh the point meant "I want this scarf," so the polite dolphin thing to do was to take the other. Fourth, maybe she was mimicking my point. All are possible explanations and none are clearly obvious. I was used to wild dolphins and I felt it was more worthwhile to just expose them to a communication system and see where they led us. I wanted to get back to the side of the work designed to give the dolphins time to explore this system spontaneously. Realizing that setting up experiments in the water was too difficult, we continued the modeling and demonstrations of the system as our primary method.

It was now mid-August and it had been a slow trip for dolphins at least during the day. One night we found a group of spotted dolphins to the south of the nursery, and set up for a night drift and watched them for four hours as they fed in the deep. It was always amazing to watch them at the edge, drifting along using their caloric expenditure for the capture of flying fish and squid. The fact that they spent hours and evenings foraging on the edge enlarged our vision of their lives. We saw a group of young adult males, Flash, Cat Scratches, Punchy, and Big Wave fishing. Surprisingly, Heaven, a juvenile female, was with these males. It was a bit surprising seeing a three-year-old female with this group. Was it unusual for young dolphins to be on the edge at night? It was a great place to practice hunting skills, but the fish did get away sometimes. The fish played dead, floated, hid among clumps of sargassum, or, if we were in the water, hid in our hair and cameras. They also hid against the boat and the dolphins blew bubbles underneath the fish to dislodge them, a handy trick. Dolphins chased live fish and avoided injured or strange fish. When a few flying fish hit the side of the boat or landed on deck, stunned, we threw them back in the water. The floating fish appeared an easy grab for the dolphins, but the dolphins hesitated, mouthed the fish, dropped it or watched it, but never ate it.

Later that week we encountered a large group of dolphins. Caroh initiated play with a piece of sargassum, teaching her friend Lava how to get people to play the game. Dr. Fabienne Delfour, my French colleague and also a cognitive scientist helping with the research, and I played with the scarf with Caroh and Lava and then, as we had come to expect, Caroh swam away with the red scarf (number seven). By the end of the summer Caroh and some of the other dolphins had begun to station themselves in front of the keyboard and oriented to specific keys. Taking their positioning as an intention to point, we pulled out the scarf or rope, depending on which symbol they oriented on. We played our game, interjecting ourselves at the keyboard when we wanted another human's or dolphin's toy in play. Games went on for hours and throughout our sessions we used the keyboard to request or change toys, further demonstrating the power of this tool. Fabienne and I counted and tracked all the types of exposures each individual dolphin received. How many times was Caroh around when we used the keyboard with the scarf symbol? How many times did Tink take the lead during the keyboard work when Caroh was there? It was clear that these main ambassador dolphins had their own dynamic and we felt it critical to track the interactions both with us and the other dolphins.

Our research trips were from ten days to three weeks long, giving us a chance to really focus on the work and build up our team dynamic. In between trips we crossed to Florida and watched pantropical spotted dolphins and offshore bottlenose dolphins traveling in the Gulf Stream in their deepwater habitat. To break the monotony of three weeks at sea we went to an island or reef for a day or two. In the middle of these long stretches we had "formal night," which involved dressing up for dinner, in togas, in ties, whatever was different. We lit the torches and put music on. We played games, especially charades, always a favorite. At the end of three weeks we often ran out of green produce and often begged or traded for a head of lettuce with a canned good off a local dive boat. Cooking could be entertaining, too, and we all pitched in and cooked a meal, giving our regular cook a break. We were blessed by good weather for the first two years of the two-way work and had great teams of humans, skilled captains and first mates who kept the boat running and the water-maker working, and of course the dolphins, who kept the work exciting and our creative levels high. For the first time there were no hurricane threats at the end of the season, which helped ease us back into our world on land back in Florida.

The possibility of communicating with nonhuman animals has fascinated humans as early as Aristotle. Interspecies interaction, specifically human-dolphin interaction, is a very old phenomenon, but only recently a field of scientific inquiry. Stories about dolphins befriending people continue to be reported, including rescuing swimmers, guiding lost vessels, and aiding fishermen in their catch.

Perhaps because of expediency, control, and access, the most serious efforts for interspecies interaction have occurred in laboratory settings.

The current paradigms of science barely allow for the emergence of interspecies investigation. Interaction with dolphins is a particularly sensitive issue to many who would shun the concept of a nonhuman intelligence. Although studies of captive species can be productive for some aspects of cognitive and language work, I believe that a species is not studied to its full potential if studied only outside its natural environment. A dolphin is only a dolphin in its ocean habitat surrounded by members of its own society and dealing with the everyday issues of dolphin life such as foraging, mating, and socializing. I also believe that not every dolphin on the planet is curious or has an interest in interacting with us. Humans are unique individuals with personalities and preferences. The same is likely true for many other species including dolphins. Disturbing the natural lives of dolphins by assuming they would like to interact with us is as dangerous as taking them out of the wild. It does not acknowledge their presence as individual beings with lives of their own and the integrity to choose when and with whom to interact. I believe it is impossible to truly understand the essence of dolphins unless our participation with them is mutually voluntary and respectful. I think this is why many people are drawn to dolphins. Dolphins represent a healthy and live connection with the Earth and one that has the potential to re-create our place and relationship with nature.

The human history of interacting with dolphins is less than pure. It includes consumptive uses of cetaceans (whales and dolphins); incidental catches in the tuna industry; captive educational, research, and New Age therapy facilities; captive swim programs; and open sea interspecies interaction. While the latter activity is the most benign, it is certainly not the most frequent type of contact. But on whose terms are we interacting? At what point is it voluntary and desirable by both species? Our search for other intelligent minds has led us to the cetaceans as well as other large-brained, social mammals. We may find universal patterns of intelligence, communication, and mind that show continuity through evolution and between aquatic and terrestrial species. But how intelligent are we willing to let them be?

If we really want to go down the road of interspecies communication there are a few promising directions we could explore with dolphins and other species. Most important, a species must *want* to communicate. The familiarity with and acknowledgment of the importance of social bonds to the development of a mutual language cannot be underestimated. Decades ago Ken Norris compared observing nonhuman animal cultures with work done by the cultural anthropologists Margaret Mead and Gregory Bateson. He speculated that animal societies have rituals and ceremonies that help transmit and sustain the culture. To date, four species of cetaceans—killer whales, bottlenose dolphins, humpback whales, and sperm whales—show aspects that favor social processes conducive toward culture. My own observations suggest that both spotted and bottlenose dolphins in the Bahamas also have cultures. My guess is that most small dolphin species have cultures, but they have yet to be systematically analyzed.

First, most cultures transmit both specific and complex information, so a natural place to start would be to work very hard, with cutting-edge technology and software, to *decipher* their communication code. This not only gives us information about the complexity of their society, but it could yield valuable information about how to design a two-way system. Other species, including many sentinel species, already understand and tap into their neighbor's communication signals to receive warnings or predator information. There is probably valuable information in an animal's own communication system currently undeciphered.

Second we could look for *universal features* across communication systems that might give us a larger "meta" system for communication. Eugene Morton's research has already shown that there are both structural and motivational rules, based on the harshness and the frequency of sound, that many birds and mammals follow. Patricia McConnell's research on the similarities of whistles used by humans to communicate with dogs found they have cross-cultural patterns that naturally emerge.

Third, a *shared, artificial communication* has the ability to open a window to interaction with other species. Flexibly designed, a two-way system could both incorporate natural communication signals and mutually agreed-on, but artificial, signals that both participant species might utilize. In World War II courting Japanese and American couples created their own language to bridge the gap of communication. Continuity of communication can be found throughout our sensory systems, including the coding of facial expressions eloquently described by Paul Ekman and other researchers. "Synesthesia," the term for cross-sensory perception, suggests not only that humans may have the ability to hear color and taste music, but that natural translation mechanisms in our sensory systems to allow understanding already exist. Studies show similar chemical and neurobiological systems between species, the internal substrate for emotional continuity. Centuries ago, Charles Darwin described his observations and ideas of evolutionary continuity, including six universal emotions: anger, happiness, sadness, disgust, fear, and surprise. Other researchers have added jealousy, contempt, shame, and embarrassment. Antonio Damasio, in his book *Descartes' Error,* adds other social emotions including sympathy, guilt, pride, envy, admiration, and indignation to the list.

We see endearing emotional relationships forming between members of the same species and sometimes improbable relationships between animals of wildly different species. Carnivores and powerful predators can also show compassion, sympathy, and empathy. Humpback whales have been observed defending other species from orca attacks. Like young prespeech human children, nonhuman animals probably know more than we give them credit for, but just don't have the tools for expression. Or it may simply be that humans do not have the tools to translate.

To a certain extent humans may have a natural ability to recognize emotional expression in other species. The context of signals in social mammals may have some overlap. A young dolphin anxiously and excitedly greets its returning mother and the threat and anger in competing males is noisy and escalated, examples that show a surprising parallel to the emotional signature of human bonding and fighting. The modalities may be different, but the message may be the same. Social learning, and specifically cross-species social learning in the wild, could provide a new step down the path toward interspecies communication. And there lies the possibility of universal language.

1999—Keyboards and Communication

1997 and 1998 had been incredible years for the two-way work with a group of female juvenile spotted dolphins that became our star players. We continued the Phase II work in 1999 with a few exceptional moments, one representing our first true test of whether the dolphins understood the labels on the keyboard.

Our normal season started as usual with updates of spot patterns and new calves of the season. May had been plagued by swarms of thimble moon jellyfish in our area, turning our clear water into a gelatinous swimming experience. Sea lice (larval jellyfish) attacked us relentlessly, sometimes chasing us out of the water with measleslike bumps as souvenirs. Large tiger sharks were looming nearby on the sandbank where some of the U.S. dive boats had started to chum for sharks, and in the Gulf Stream pods of pantropical spotted dolphins and pilot whales traveled north with their families to some unknown location.

By early July we were once again focusing on the two-way work. One day while Caroh and Adam were playing with a scarf and the keyboard, a bottlenose dolphin came into the encounter and interacted. But Caroh wouldn't give up the scarf (except to her friend Tink) and the bottlenose drifted away. We chased and played while Caroh used the scarf to bump and nudge us. In the end we couldn't get the scarf back from Caroh and as a hammerhead swam under us, forcing us to retreat to the boat, Caroh swam away with scarf number nine.

Through the summer we used our keyboard system with the northern pod, including Paint, her daughter Brush, and friends. Little Gash, Big Wave, and Laguna all explored the jars when they were on the anchor line, and the keyboard became more and more of an object of interest to an expanding list of dolphins. By early August Caroh had taken off with scarf number eleven and we were rapidly running out of scarves. But then one incredible session with the dolphins occurred that showed us that perhaps they did understand the different symbols on the keyboard and that, after three years of work together, we understood each other.

We were working with the keyboard in relatively deep water, forty-five feet or so, which limited our abilities, and using snorkels to retrieve objects that we wanted to exchange with the dolphins. Caroh and Tink slowly swam over to the keyboard and hovered vertically as they watched us demonstrate the keys and exchange toys with our fellow humans. Tink had the rope toy and she dove down deep. She and her fellow dolphins played back and forth with the rope toy on the bottom. I saw that this was the perfect situation to request the toy from her using the keyboard. I swam over to the keyboard that was in the water nearby, and triggered the sound requesting the rope, and Tink immediately ascended from the depths, rope in mouth, and dropped it right in front of us. She had come up from the bottom with the rope in her mouth and had given us the toy. We clapped our hands and yelled excitedly through our snorkels. The excitement was contagious and palpable. The dolphins rapidly rubbed their pectoral fins together, but kept us in view from behind as they often did when they were excited. It was the first time I had requested a toy using the keyboard and received such a clear response. It was a beginning at least, but one we needed to test more, to confirm whether the dolphins were indeed discriminating from the different symbols and toys their activation provided. I remembered the past weeks when our slew of volunteer cooks and colleagues pushed and swam the equipment encounter after encounter in heavy currents to keep up with the dolphins. It had all paid off. This, to me, was the culmination of all those sessions. We had opened a window at least. But where could we go from here? I wondered where we could take this communication. Right now the dolphins or humans could request a toy using the technology. There was no doubt that the technology was slow and too simple for the dolphins. By 1999 I had hoped to have a computer in the water, human and dolphin friendly, to share complicated sounds and contingencies, and ready at a day's notice to add new vocabulary to the equipment to keep up with the dolphins' fast pace. I had no doubt that with the right technology we could take communication between our two species to great levels. Ironically the technology existed, the field site existed, but as with most projects, securing the funding to design, build, and implement a project was in short supply. In addition, the dolphins had been ready and willing over the last four years, but they, too, were moving on in their lives. We were losing our pool of available dolphins since our juvenile females were coming into sexual maturity and other responsibilities. At a minimum I could continue to decipher the dolphins' communication signals and plan for the next system to implement in the future. I was determined to search out and develop better hardware and software to work in the water in the future. We needed a more advanced and dolphin-friendly system to use in the water. We needed real-time sound recognition that was quick, needed a computer in the water with hydrophones, and an English translator for us nondolphin types. Even after securing a bit of funding for this work it just wasn't enough and the development of these toys died a slow death.

What really stood out in these extraordinary years of two-way work were the importance of individual dolphins and personalities and the right team of humans helping with the work. Some dolphins, like Caroh, had many hours of exposure to the keyboard. Other dolphins remained on the periphery, never becoming major players. Yet other dolphins, such as Tink and Mitsu, at first only observed, but then gradually took initiative and explored the communication system. Some dolphins were truly ambassadors with humans while others chose to stay away. Juvenile female spotted dolphins were our best candidates for this work. This is the age class when spotted dolphins are newly independent, but with at least a little dolphin wisdom. Females are a bit more fluid in their relationships than young males who are already developing their male coalitions and fighting at this age. It was also striking that some days the same dolphins would come back three or four times to work with Phase II equipment. Their interest was intense and the weather held for us many weeks at a time, enabling us to remain anchored in one location so the dolphins could find us.

It was also clear that the dolphins had the interest and that we had illuminated a few critical aspects of this work. I think this species is perhaps the most intelligent species on the planet after humans and they naturally seek out stimulation and interaction. Intelligence seeks intelligence. There are a few other dolphin communities around the world that have this potential for the future. As September approached, and Hurricane Floyd threatened south Florida briefly before veering off as it bounced off the Bahamas, I dwelt on the need for dolphin-friendly technology to take our communication further and wondered how I might find it.

~ 8 ~

Advancing the Work

Man can learn nothing unless he proceeds from the known to the unknown.

—Claude Bernard, 1928

2000—Losing Lanai

It was the beginning of the new millennium, but it felt like the end of Phase II. Last year had been spectacular, ending in a great session of keyboard use by Tink and her friends. I tried a few sessions this summer, but I desperately needed better dolphin-friendly equipment and our window was closing with the current group of available dolphins. As we had seen over the last three years, juvenile females were our main players, but Caroh, Tink, and Rosepetal were all rapidly maturing and entering their reproductive years. It was unlikely they would be active in a two-way system given their new responsibilities. Frustrated, but equally encouraged by our success, I decided to take a break from the two-way work until I could put together better technology and more funding for the next attempt. So I went in search of money, colleagues, and technology that might propel us forward into our next attempt at two-way communication.

But our field season continued as it normally did and we watched dolphin behavior and monitored the population, including the new calves. Last year I had noticed that Rosemole, Rosebud, and White Patches were all missing from the group. Occasionally an individual dolphin is not seen one season but shows up the next year. In our census log we give them three years before we list them as lost. But it was disturbing because these three females were always around in a normal season, and last year I had noticed the void painfully. I watched anxiously through the summer but with no luck, Rosemole, her first calf Rosebud, and White Patches were nowhere to be seen. I felt the deep loss of some of my first dolphin friends that I had met fifteen years ago.

Life and death are natural cycles in any society and the birth of a calf is therefore a special and precious event. Although calving peaks are traditionally in the spring and fall, occasionally we bear witness to calving activity in the late summer. The loss of a calf represents a devastating moment in a dolphin's life. In 2000, I observed an amazing episode between a spotted dolphin, Luna, and her newborn calf, Lanai. I've known Luna since 1985 when she was at least fifteen years old and more likely twenty years of age. I was able to estimate her age because the spots on her body were already fused and coalesced black and white marks. Like other females her age, Luna was an experienced mother and we had observed her with four different offspring over the years. Giving birth every three or four years is typical in this society and Luna was usually on schedule.

One hot summer day on the sandbanks in May 1999 my research team and I were out observing our dolphins as dark gray squalls formed in the distance over a calm sea. I watched from the surface as some of the dolphins foraged under frigate birds for their lunch. Then the dolphins began grouping up and performing synchronized leaps from the water. I decided to get a closer look and entered the water. I saw a large male coalition with Luna as the central female. Luna was doing aerial head-over-tail flips and side breaches. Rivet, a maturing male, was challenging older males Horseshoe and White Spot, as is typical during courtship activity. Luna remained "herded" by the group, but continued doing aerial leaps. Shortly after this encounter we moved our research vessel and picked up Venus and her calf Violet on the bow. Two older males were accompanying these young females. It was clear from the males' markings that they were the same ones we'd been seeing courting females—especially Luna and Venus—all spring.

Luna was often escorted by Rivet, an aggressive and up-and-coming male in the community. One day I observed Rivet performing some aggressive postures, including arching his body and hanging at the surface motionless. This is a protective gesture that males often perform while escorting pregnant females. Luna brushed Rivet physically once, but he ignored her and kept an eye on us. Since dolphins have a twelve-month gestation period, we expected that we would see Luna's new calf when we returned the following May. But when we began our seasonal observations in May 2000, there was no new calf and Luna appeared to be about six months pregnant. I could only deduce that Luna must have miscarried sometime in the summer of 1999 and then gone into postpartum estrus and conceived again.

Dolphins can appear very pregnant even as they continue to enlarge over the months, seemingly ready to burst with new life in front of us. On August 11, 2000, I observed Luna on the sandbank in her very pregnant state. Four days later, on August 14, I observed Luna again, this time with a newborn calf. The calf had fetal folds and a floppy dorsal fin waving in the water and swam awkwardly, as newborns do. As we approached we could see this very small calf (who we ended up naming Lanai) nursing. There were other mother-calf pairs around, as well as some old males. The dolphins were on the move with their school of hungry little tunny in tow, likely waiting opportunistically for a snack chased up by the dolphins. As the group traveled slowly to the southwest we had only a brief glimpse of the calf's features. This was indeed a first. I had the rare opportunity to document the birth of a wild dolphin within a four-day period.

A few days after we saw Luna and her new calf I decided to take our boat farther north, to an area that we survey only occasionally. Since we were already at the northern border of our study area, I wanted to make use of a few days of good weather to search for other dolphins. For years I had heard, from the local Bahamians, about a group of spotted dolphins to the north. Only once in all our years of surveys did we chance upon this group, and only for a brief period and in deep water at twilight. Who were these spotted dolphins? Did they overlap with our study group? Is this where our immigrant dolphins, usually juveniles, came from every year? To answer these questions we headed up north to Walker's Cay, near Matanilla reef, the place where locals told us these spotted dolphins lived.

A healthy and vibrant reef on the northwest corner of Little Bahama Bank, Matanilla is notorious for its large breaking waves and large tiger sharks. Though it is not a safe area for either dolphins or researchers, we decided to explore it anyway. After a day without any sightings we headed back down to our normal study area. On the way south we saw a large group of dolphins, but something was odd. It was a large group of bottlenose dolphins, ten or more, with one lone spotted dolphin. Although we frequently see these two species together, usually the ratio is reversed—more spotted dolphins than bottlenose dolphins. This piqued my interest and I decided to get in the water to see what was happening.

When I entered the water I noticed two things straightaway. First, the water was very murky, which happens during low tidal cycles, but I could still see the spotted dolphin, hanging stationary in the middle of this large group of bottlenose dolphins. Secondly, I could hear an underwater cacophony of intense vocalizations and over all this background noise I could make out a distinct, frantic, and familiar whistle correlated with a bubble trail coming out of the spotted dolphin's blowhole. It was Luna. And I recognized her signature whistle. It took a while for me to register the whole scene, but as I absorbed what I was seeing it occurred to me: "Luna is here without her newborn calf. She sounds frantic and is making her whistle, as if to call her calf back." I immediately thought, "Luna, distress, signature whistle, no calf. . . . Is there a shark around? Is Luna frantically calling for her calf shortly after it was lost to a shark?"

It took a few more moments for my brain to realize the potential danger we, the researchers, were in. Just at that moment there was a tug on my flipper. I turned around to see my assistant Sara in the murky water with a large, ten-foot tiger shark swimming leisurely behind her. We both popped to the surface to talk. "There is a large tiger shark right behind us," Sara said in a calm but concerned voice.

There is nothing as primal as the fear evoked by seeing a large shark in the water. Especially one focused on you. Your scientifically trained brain can register that it's a beautiful animal, so important in its own ecosystem, but your primate brain is saying "run"—or in this case "swim"—and get out of the water where we humans will be a bit safer. In fact diving in the Bahamas is quite safe and rarely do sharks bite divers (other than when humans spear fish or if the sharks have become accustomed to regular feedings from local dive boats). Tiger sharks, although notorious man-eaters in the Pacific, have a different reputation in our observation area. In the Bahamas they are both timid and shy. However, they are still large, often hungry animals and we take safety very seriously while doing our research.

Sara and I calmly waved the boat over, as our safety guidelines dictate, while keeping a watchful eye on the tiger shark on our heels. We emerged from the water, intact and a bit scared, but also frustrated because we had to cut short such an interesting observation. We explained to the crew the strange circumstances we had just observed and decided to stay with this group of dolphins to observe them from the surface. Unfortunately, we managed to lose this large group of dolphins. Bottlenose dolphins in our study area are notorious for disappearing on us and we jokingly refer to them as Houdini fish for their disappearing acts. Frustrated, we continued heading south to our regular study site nearby.

We arrived at our normal study site and our mystery got a fresh twist. Lanai, Luna's four-day-old calf, suddenly appeared at the bow with Caroh and Tink and Lanai was trying desperately to suckle from Tink. Why wasn't Luna here with Lanai? How did Lanai get here with this group? Could she possibly survive without her mother's milk?

Many scenarios crossed my mind. Did this group of juvenile female dolphins swim off with the vulnerable calf, while Luna and the bottlenose dolphins fought a tiger shark? Did Caroh or Tink "steal" the calf from her mother during a shark encounter? Female dolphins, usually before they have their own offspring, have been known to kidnap young calves from mothers, at least in captivity. Kidnapping behavior has also been observed in chimpanzees in the wild as well as in captivity. Could this be the first observation of a dolphin kidnapping in the wild? Or did the calf simply swim away with other members of the pod for her safety during a shark attack?

In all the time I had been in the field with this community of dolphins I had never heard a mother more vocally distressed nor witnessed such a confusing interspecies scenario. Maybe Luna's stress was due to not knowing where her calf was or if it was still alive. The story took an even odder turn: as I examined Luna's newborn calf closely for the first time it seemed to me as if she might be a hybrid between a spotted and bottlenose dolphin. Is it possible the group of bottlenose dolphins surrounding Luna recognized the calf as a hybrid? Hybrids can sometimes be identified by the combination of morphological features from two different species. Lanai was certainly another good hybrid candidate.

We never saw Luna with Lanai again. Lanai, less than a week old, likely did not survive without her mother. Our team was haunted by the question of what happened to our old friend Luna. At the very end of our field season in 2000 we observed a large group of twenty-five spotted dolphins and then ran into a second group of dolphins that were traveling and chasing fish under frigate birds. I thought I saw Luna from the surface (without Lanai and with a large group of adult dolphins), but I only had her dorsal fin shape to use for identification. I went into the water in time to see some male coalitions swim by, but they were very preoccupied foraging on the sandbank. I was disappointed and thought that I simply would never know the end of the story. I have never observed a dead dolphin in our study site. There are no nearby shorelines for their bodies to come ashore. If a dolphin died in the open ocean, it would likely sink or get eaten fairly quickly.

Could a dolphin actually die of grief? In her book *In the Shadow of Man,* Jane Goodall describes the grief, despondence, and death of a male chimp after his mother died. We know that with humans, grief and strong emotions can affect our immune systems and our health. A decade ago I had witnessed Gemini, despondent from her loss of Gemer. I had observed Mugsy swimming listlessly after the loss of her first calf. Luna could have suffered similar shock and depression at the loss of her calf. Could Luna actually die from grief over her lost calf after giving birth so many times before? The odds of survival for any dolphin are mixed. Despite Luna's success at reproducing at regular intervals, and successfully weaning most of her offspring, only one offspring, Latitude, still survives. Latitude is now a fully grown male, vivacious and filling the role that male dolphins do in their society. He has a regular male coalition with whom he hunts, courts females, and chases bottlenose dolphins. To my knowledge, Latitude is the only offspring to carry on the genes of his mother Luna.

I work in the best place in the world to observe dolphin behavior underwater. I have spent about four months every summer for twenty-five years, day after day, for more than two thousand hours, observing dolphins underwater. Even with this much observation and accrued knowledge of individual dolphins, families, and relationships, putting the pieces together is still challenging. For example, female aggression occasionally flares up, but it is much more subtle than male aggression. Females, in the course of soliciting the attention of males, gently engage in head-to-head fights. It was rare that females ganged up on each other, they just didn't have coalitions, but instead a broad network of friends. Sometimes they would start something and end it just as quick, as if they were reading subtle signals of negotiation barely detectable (at least by me). It was rare, therefore, for us to record many examples of female fighting. In Luna's case I was also very lucky to see the bits and pieces preceding, and following, the unique event surrounding the birth of Lanai. The possible scenarios, and their potential implications, are endless. I can speculate. I can imagine reasonable scenarios based on what I know of these dolphin's lives. I can give it my best guess and even statistically analyze these events. But what remains is, and will likely always be, a mystery.

2001—A Sound Society

Through the winter months I had been writing grants and fund-raising, but I was anxious to get back out into the field. Although the two-way work was on hold for a while I knew the importance of being in the field every summer and tracking the identification marks of the new calves and regular dolphins—it was the framework for all our work. Even though many observations were routine, there was always something new to observe about the dolphins' behavior.

Mother's Day heralded in not only the normal bloom of sea lice, but some spectacular weather. A long-term friend of the project, Ruth Petzold, was out with us to help with our underwater photography of the dolphins' spot patterns. An avid ocean explorer, Ruth had recently lost her leg to an infection, but her intrepid spirit couldn't stop her interest in the water world. As she floated in the water with a mother and calf, we wondered if they would notice that this human was missing an appendage. After all, they lose chunks of their flukes and fins over time. But as Ruth focused her camera on the new calf they swam quietly around her, quite a different reaction from the exploratory echolocation clicks we had expected to hear as they approached her. It somehow seemed appropriate on Mother's Day.

This summer was also the first time we observed the dolphins balling up fish. We were watching a large surface-active group and when we entered the water we noticed the dolphins corralling a school of bar jacks. Simultaneously the bottlenose dolphins present were chasing and mounting the younger spotted dolphins. It was an interspecies picnic. As I tried to videotape the male activity, the bar jacks schooled around the camera trying to escape the occasional chases of the spotted dolphins. To my left I could see Chris Traughber blending into the fish ball that now surrounded him as he photographed the activity and eased away from the fish. There were bottlenose and spotted dolphin fights, spotted and spotted dolphin fights, and a foraging frenzy all happening at once. Only once did I see a fish caught and eaten. Later that day we found the same group of dolphins still corralling a fish ball. This time we stayed back farther from the action, trying to discourage the frightened fish from taking shelter among us. Perhaps the dolphins were corralling fish all these years and we never observed it. Perhaps their prey sources had changed and they were using the schooling fish more than the bottom fish we often saw them eat. The next day we found the same group, still with their precious fish ball. We had been following sixty spotted dolphins when they stopped near the edge of the sandbank and were joined by four bottlenose dolphins. The younger dolphins and mothers and calves were balling up the bar jacks, dashing in and out, grabbing and swallowing them. Suddenly a very pregnant bottlenose dolphin darted into the ball, screeched to a halt, whacked a fish with her tail, and then did it again, apparently showing her own style of hunting. We watched as the dolphins, still corralling what appeared to be their portable picnic lunch, traveled on.

It was becoming clear that the dolphins had many different ways of learning. As newborns, calves stay underneath their mother and listen and watch while fish are caught. Later in the calf's life a mother, or older juvenile, encourages the youngster to chase a fish on the bottom. Sometimes groups of young male dolphins imitate their elders on the periphery of an escalated fight. Sometimes it was pure observation, other times trial and error, and sometimes it involved teaching. One thing was certain, dolphins were master mimics, acoustically and posturally, and they used these skills in everyday life.

Without knowing the relationship of interacting dolphins and the dynamics of their society, it would be difficult to interpret signals as friendly or aggressive. In a fully actuated healthy dolphin society that is running smoothly, emotions can be nuanced. Changes in posture, eye movement, and vocalization rates can all change slightly. Play is one way to practice social behaviors, including courtship, mating, and fighting, without the severe consequences of actually being engaged in the real thing. Play behaviors have all the elements or "events" found in escalated adult behavior, but at different rates and intensities. Immature dolphins learn the ground rules during play in a nonthreatening environment. Female styles may also vary and so does the training of their calves. Jane Goodall has described the different mothering styles of female chimpanzees in her work. Longitudinal research allows comparisons across generations and begins to illuminate the diversity of behavior and variations of behavior found in complex societies.

A month later I watched Pigment, Paint's offspring, orient her body in a point and show Sunami, Snow's offspring, how to catch flounder and snakefish on the bottom. As the little tunny followed the dolphins, snagging the occasional flounder left behind, I wondered if the little tunny were occasionally a meal for the spotted dolphins, yet another mobile picnic lunch. Over the years I have never seen a spotted dolphin eat a little tunny, but I have seen the dolphins get annoyed at a fish. After becoming overly annoyed at a little tunny, Dash, a young male spotted dolphin, killed one and raked the fish thoroughly with teeth marks, shaking it and dropping it as he swam away. I still have the rake-marked little tunny preserved in a jar, a victim of his own attempt to steal Dash's fish.

There were still so many basic things we didn't know about dolphin vocalizations, and anything I could learn about their sounds might help the future of our two-way work. How much of their sound was ultrasonic? Did their whistles have harmonics above our hearing? We suspected that dolphins made sounds above our hearing range, at least echolocation clicks, that span broad frequencies, but we just didn't have a good portable system for our field site. Years earlier, colleague Kelly Allman had joined me with high-frequency hydrophones and a tape-recording system. But the dolphins were often distant from the boat when they were doing the behaviors, such as fighting and courting, that we wanted to correlate with ultrasounds. By this time there were new field technologies related to high-frequency sound collection. Doctoral student Marc Lammers had been working in Hawaii with spinner dolphins and had created some hardware and software to collect ultrasonic sounds. So, with the help of a grant for his nonprofit, Marc created two mobile collection systems, Underwater Dolphin Data Acquisition System (UDDAS). We fondly called it BADDAS on the boat.

One unit was for the spinner dolphins in Hawaii, named Spin, and the other was for the spotted dolphins in the Bahamas, affectionately known as Spot. So Spin and Spot went to work. Although fairly compact, Spot was a bit harder than my normal video to push through the water while following an active group of dolphins. When I anticipated a whistle I triggered the unit to record the dolphin whistling. Since dolphin sounds are very directional we sampled dolphins as they were swimming toward or away from the unit and its directional hydrophone. In the scientific literature it is likely that the cataloging of vocalization types represent only partial sounds, cut off from recording as the dolphin changed angle underwater, either toward or away from the recording hydrophone. For example, signature whistles often have multicontoured loops, and researchers struggle how to categorize them, because sometimes the whistles don't look continuous; their tops are cut off like a giant thunderhead in the summer sky that rips apart from the high level winds. I suspect we will revamp our idea of what the repertoire of a dolphin is after recording full-bandwidth signals, head-on, with visual confirmation.

It was the first day of August and the weather was growing worse and worse. We decided to run down to Sandy Cay to take shelter, but who should show up but Stubby, Horseshoe, and White Spot. Anxious to try to coordinate our two high-frequency units, Marc got in with Spin, and I with Spot. The southern females were there and as Venus bumped Marc and I alternately, I noted that Flying A, Infinity, and Java were all very pregnant, not surprising for late summer. As I recorded some high-frequency samples of Venus whistling nearby, Marc grabbed some other sounds and Adam got a fecal sample from Horseshoe. It was a productive encounter and, as many are, multipurpose. We exited the water with both sound samples and potential genetic material. Some of the whistles had harmonics well above our hearing range. Most of the clicks and burst-pulsed squawks were more than a hundred kilohertz in frequency, as we suspected they might be. It was a good day.

To date, most of the sound types we had recorded previously in our narrowband system (human audible frequencies) showed information above our hearing range with our new system. In the case of whistles, although the main contour was visible with our old equipment, now we saw harmonics up to seventy, ninety, or a hundred kilohertz (five times our hearing). Even burst-pulsed sounds and clicks had energy in the ultrasonic range. There were times when there was no evidence of a signal in narrowband, but all the energy was above our hearing range. So we were missing some signals, but how often? This is one reason that I began trying to sample during "quiet" behavior. I grabbed samples when the water was murky and the dolphins were swimming quietly in case they were making ultrasonic vocalizations (they weren't). I grabbed samples when the dolphins were scanning the bottom from forty feet up, but we heard no sounds, suspecting that either they were using ultrasound or just listening (still working on that one). I grabbed samples from a group of males, who looked like they were fighting with head-to-heads and open-mouth postures but no audible squawks or screams, to see about their use of ultrasound in these circumstances (no sounds were recorded). This is the reality of studying any animal, most of which have sensory systems beyond or different than ours. We must be aware of this and seek to find technologies that help us "see" like they do. All sorts of cues abound in the natural world: the ultraviolet light used by bees, the odiferous perfumery used by dogs, and the electrical eels' shock signals. To understand wild animals we must strive to get into their world.

Sometimes the dolphin's behavior seemed simple and basic and other times it seemed overwhelmingly complex. How do you really look at complex behavior in an intelligent species? Although some might argue that animals in the wild have little to communicate about, it would seem that their daily challenges and relationships would point to a myriad of needs to communicate both emotional and contextual information. Just managing the complex politics of brothers, sisters, coalition, and neighbors seems need enough to have long-term memory and behavioral subtleties.

Correlating behavior with vocalizations is one way to start to understand the complex communication among dolphins. We had, for more than a decade, correlated basic sound types with behavior: squawks and screams during fighting, signature whistles during mother-calf reunions, echolocation clicks during foraging behavior, genital buzzes during courtship activity. Most important to remember is that behavior is contextual. I can watch a mother swimming upside down and inverted after a calf as an indication of discipline, and I can see the same upside-down posture when a male is chasing a female to mate. The important thing is that I know the players, their age, sex, and history. I can see what transpires during the interaction; the calf is held down on the bottom in discipline, or copulation between the male and female. Like Jane Goodall's description of six different chimpanzee courtship gestures observed in various sexual contexts, such subtle variations, potentially ascribing different meanings, may be analogous to the multiple functions we see in the dolphin tail slap (attention, annoyance, emphasis), and inverted chases (discipline, mating). It is the constellation of factors that are recognized and negotiated by individuals in the society that make up a communication system. Communication signals are multimodal and include body postures, types of touch, eye movements, body movements, and angles of orientation, and all play into the process of communication in a complex species.

Because of a dolphin's sophisticated acoustic abilities, communication research has often focused on the vocalizations of dolphins. Do dolphins have a language? Do they have words (referential signals) or just express emotional states (graded signals)? Although signature whistles may fit the idea of a word or reference, since they are referencing a specific individual, there are many other aspects that can be critical in a communication system. Prosodics, which include the rhythm and intonation of signals, is one of them. In humans it is a natural tendency for adults to adjust their speech to infants, including speaking at a slower rate, increasing the frequency of sounds, and emphasizing rhythm and intonation. Developmentally, rhythm and intonation emerge before words, suggesting that such a process may be aided by synchrony and communication skills while growing up. Primate vocalizations such as the chimpanzee "pant-hoot," described by Christophe Boesch and colleagues, are uttered with voice intonation and rhythm, suggesting that these calls may be sensitive to the real-time shifts of social interaction.

Rarely, however, have researchers analyzed the sequence of signals. Is it in the escalation of such signals that meaning is found? If your voice gets louder and louder in an argument does this mean something between you and your interacting partner? During dolphin fights the escalation of aggression shows an increasing number of signals, and loudness also increases as the situation reaches a crescendo. Analyzing sequences may be as important, if not more so, than the structure of an isolated whistle or click. Measuring signals, whether they are sounds or body postures, in their order of increasing intensity or subtlety, and then noting the subsequent behavior, may be critical.

Another aspect of communication is the "audience" effect. *Who* is there, potentially watching or interacting, may affect the signals that an individual produces. Is there an estrous female watching or a dominant male? Is the male a coalition partner or someone who has historically been antagonistic? Relationships can at times be affiliative, antagonistic, or neutral, changing with the audience. It should not be surprising that dolphins, like other animals, may show real-time negotiations. At a distance I can hear the screams and whistles and know that a group of dolphins is fighting. Likewise the vocalizing dolphin's ongoing sound and behavior is heard and acted upon by other dolphins, some rushing to the scene, others fleeing. It is rare that a dolphin doesn't have an audience.

Two large issues dominate when trying to understand dolphin acoustic communication. One is the difficulty in determining who is vocalizing and the other is the accurate recording of the directionality of the sounds. Dolphins produce (and hear) human-audible and ultrasonic vocalizations that are extremely directional, and the higher the frequency the more directional they are. Whether the researcher acquires the harmonics depends on the orientation of the dolphin to the equipment, and of course the equipment itself. My personal belief is that we have barely scratched the surface of dolphin acoustic signals because historically researchers have not been in control of the orientation of their hydrophones, or the dolphins, while recording dolphin sounds (like hanging a hydrophone over the side of a boat). In my own research site, probably the best in the world to observe underwater interaction, I can see which dolphin is orienting to my hydrophone and, even with a twenty-five-year database of individuals and relationships, the entirety of these signals are still difficult to obtain. When a dolphin is alone and in close proximity to my underwater video and hydrophone, and when there are bubbles coming out of the blowhole, I am sure of who is whistling. In many situations dolphins are in groups or do not make bubbles when vocalizing.

Dolphins do things that are amazing and sometimes frustrating to the researcher, including internally directing sound at an angle not indicated by the position of their head. They have two nasal passages, which merge into one blowhole, and can make two sounds at once, and they can make sounds above our hearing range. The majority of our underwater video contains groups of dolphins squawking and whistling at the same time. This is analogous to you sitting at the dinner table with your family and talking all at once. Are you talking about the food on the table, what kind of day you had at work, or are you expressing emotions during an argument? In the past we have believed that only humans are capable of talking about yesterday or tomorrow, a hallmark of intelligence. But are we even looking for it in other complex species?

If dolphins can recognize one another's signature whistles, and perhaps have a "voice" or specific vocal quality embedded in all their sounds, then the dolphins potentially know who is fighting whom, how things are escalating within the group, and which friends of theirs are in the group. Additionally, we don't yet know where the "individual" voice characteristics of dolphins reside. In humans, it is our timbre, the resonant qualities of our vocal chamber, that allows our friends to recognize our voice on the phone or in a crowd. Although dolphins have "signature" whistles, thought to be distinct by their fundamental frequency modulation, it is possible that dolphins have unique "vocal" qualities that reside both in their echolocation signals and their burst-pulsed sounds.

In past animal communication studies we have not been willing to consider that nonhuman animals might be intelligent and communicate complex things, but scientists have not really had the tools to analyze complex behaviors. New frameworks are emerging, including Dynamic Systems Theory (DST) and Distributed Cognition (D-Cog), both recent methods that promise new and complex tools for the analysis of communication. In *The Dynamic Dance,* Barbara King describes the framework and specific examples of how communication signals are produced and negotiated in real time. In DST or D-Cog the process of communicating socially is not about a transfer of information, but rather the emergence of a mutual understanding through a shared action or thought. Meaning does not reside in words, but in the mutual construction of meaning among partners, created through interaction and coregulated by all participants. Increased coordination is seen between individuals as they reach meaning through negotiation. Such mutual adjustments might be seen in the adjustments chorusing male chimpanzees make to their pant-hoots in the process of call convergence. Dolphins adjust their physical movements and their vocalizations to reach physical and vocal synchrony during aggressive chases.

Meaning can exist during the interactions of even two separate species if they engage as social partners. Interspecies interactions exist in many forms and most don't include humans. Close interaction depends on careful attention to detail, knowledge of the other, and subtle adjustments in the interaction. Observing unfolding events, to another species, may just be the norm for their social world. Play is an interspecies media, an opportunity to engage and explore the other. In two species with bodily similarities such as spotted and bottlenose dolphins in the

Bahamas, the markers may be closer and more recognizable than in two disparate species.

Dolphins use mimicry and synchrony in their communication system and they are both masters of acoustic and postural mimicry. Mothers station themselves on the bow of our anchored boat during the flowing tide to train their calves how to position themselves for what later will be a real bow ride with the dangers of a moving boat. Bottlenose calves position themselves under their mothers while their heads move in synchronous rotation, presumably scanning for fish on the bottom. We watched Caroh imitate her mother as she dove down with an open mouth to stick her rostrum in the sand. We watched juvenile male dolphins mimic the aggressive signals of a group of adults nearby. We observed synchronized body postures and synchronized squawks, while male coalitions coordinated their repelling of bottlenose dolphins—clearly an expertise reached through perfect adult male execution was only an uncoordinated attempt when performed by juvenile males. Nothing could be more critical in dolphin society than the development of young dolphins and the behaviors they are exposed to by their elders. Dolphins develop their skills over time and like human children they observe adult behavior and test boundaries, and mimicry and synchrony are paramount in their learning process.

There are two striking mimicry events I remember that illustrate how the dolphins initiated activities with us. In the early years, Paint, one of the northern females, had developed her own technique of greeting me in the water. Then when Paint had her first offspring, Brush, they came by the boat. When I got in the water Paint immediately dove vertically and hung in the water column with her new calf. As these two surfaced they stationed themselves in front of me. Then the young calf went down and vertically hung by herself. Was Paint teaching Brush the signal for our greeting? Well, I wasn't going to wait to analyze it. I dove down and hung next to the calf, like I would with her mother, and Brush excitedly swam around squeaking and pec-rubbing her mom Paint. This was the Paint family tradition, and I was willing to pass it on if she was.

Years ago, during our BBC shoot, colleague Tom Fitz was down on the bottom with a scuba tank following Stubby who was hunting fish in the sand. From above I watched as Stubby made his way over to a conch shell and started pushing it with his rostrum while expelling bubbles from his blowhole that were the exact size of the bubbles coming from Tom's exhale. Stubby was imitating Tom struggle with his huge camera apparatus on the bottom and his heavy breathing.

One more incident exemplifies a dolphin's sense of humor and skill at mimicking. Humans in the water often try to "dolphin" swim, which means keeping your legs together and undulating from the waist. For the most part we are pretty bad at it, jerkily swimming while barely moving through the water. As a well-meaning passenger followed a spotted dolphin trying to get his dolphin kick right, a second dolphin followed the human, but the dolphin used jerky and awkward movements mimicking the struggling human—a dolphin mimicking a human mimicking a dolphin!

So we began to use mimicry in different situations. We waggled pieces of sargassum in front of the dolphins and they waggled pieces with their flukes or their pecs, inviting us into the game. Sometimes I tail-slapped at the surface to try to encourage the dolphins to come back to the boat; sometimes they did, sometimes they didn't (but my sense is that there are many complexities to these signals we are unaware of). Sometimes we locked elbows while swimming and the dolphins, after a double take, positioned themselves in front of us and rubbed their pectoral fins together. Sometimes I dove down to rub my hands in the sand and Paint dove down and rubbed her pectoral flippers in the sand. Sometimes when the dolphins were lying on the bottom, we dove down to lay on the bottom with them (at least when it was only twenty feet deep). If they rotated their pectoral flippers around we moved our arms and elbows around. If they did a tail stand on the bottom we tried to stand on the bottom with our fins (although being positively buoyant I tended to rise quite rapidly). Sometimes I took my snorkel out of my mouth and purposely talked in the water so bubbles were streaming from my mouth. Rosemole responded by vocalizing with exaggerated bubbles coming out of her blowhole. Were the dolphins capable of understanding an analogy? We certainly were seeing evidence of mimicry naturally and spontaneously in the wild.

After working in the open ocean in their sensory world for twenty-five years there are two things that stand out. First, dolphins are quiet most of the time, and second, they are good at picking up subtle cues of both movement and sound. We tend to think about dolphins as active producers of sound, swimming about the ocean constantly sending clicks and whistles into their environment. In the case of this community of spotted dolphins, nothing could be further from the truth. Perhaps it's because they live in crystal clear waters much of the time and have great vision. Or perhaps they are just making ultrasonic sounds that our human ears are incapable of picking up without some specialized research tools. But I don't think these are the reasons. I think that dolphins have learned and been shaped over the years to listen, exquisitely and constantly. Most animals' senses are bombarded by information. Most of our nervous system is designed to filter, block, and inhibit signals. Through the process of life we come to understand what cues are significant, and what cues are normal and insignificant background noise. I think dolphins use their passive acoustic knowledge like we constantly use our vision. Why would you need to use your echolocation to navigate if you could hear the detailed noise patterns of the waves as they slammed against the rock pile, or the current during tidal flow swirling around the reef? As our eyes pick up photons reflected off of objects, allowing us to identify and navigate between, so I imagine the dolphins listening, just listening to their environment. Swimming for hours without projecting sound certainly has its advantages; predators don't hear you, the rival spotted dolphin and the bullying bottlenose dolphins don't hear you. You are both stealth predator and stealth prey. On occasion I have observed a single dolphin, often an adult, alone in the water column and rotating her head and whistling, just whistling. It seemed that this dolphin was all at once rotating, perhaps to hear and gain a directional sense of other dolphins in the distance, and also emitting her whistle to mark her location.

Sometimes the questions that come to mind, especially when floating on the sea for months, seem bizarre. Do the dolphins see the stars? Although fully aquatic marine mammals designed for life in the water, could they tap into information above the water to navigate? It seems far-fetched. But sometimes it takes a creative experiment to test such a thought. One of my favorite examples of this is the work by Björn Mauck and colleagues on the vision of the harbor seal and whether they see the stars as potential navigational aids. They had a harbor seal look through a telescope and found he could discriminate certain intensities of stars in the night sky. In Alexandra Morton's beautiful book *Listening to Whales,* she describes killer whales in captivity observing where the first shaft of light would hit the water in their tank. Jane Goodall describes chimpanzees looking at the sunset, as if they relish the moment as much as a human might. So it seems, once again, we are reminded that creatures of the open ocean may have skills beyond what we have even dared to imagine. Cognitive complexity and the socioemotional depth of dolphins are sometimes best observed and described, as they have been in the primate literature, using anecdotes and systematic recordings of events or individuals. Until we acknowledge the socioemotional lives of nonhuman animals we will be unable to measure changes in their communication signals during empathy, grief, annoyance, and anger, a self-handicapping strategy to say the least.

Whenever I went to conferences I was always on the lookout for creative students or colleagues with cutting-edge technology. I knew enough about neural networks to know that they might be useful for our sound analysis, so I began collaborating with Dr. Volker Deecke, a creative new researcher who had developed neural network programs that essentially trained the computer to recognize a sound, say a signature whistle. Once the computer was "whistle-savvy" we could give it other whistles to compare to the now-recognized whistle. The program generates an index or measure that allows the comparison of a calf's whistle to her mother's, her siblings, and so on. After presenting this work at a conference, and showing that not only were the calves' whistles similar to their mothers' but also similar to the mother's associates, the larger family network, Peter Tyack, a renowned acoustician and whale biologist, came up to me to note that I had managed to show the larger function of the whistle in a dolphin society. Others would go on to report this in their own study sites with bottlenose dolphins including Deborah Fripp and colleagues in Sarasota Bay. So it was not just the spotted dolphins that relied on friends of the family and it was a pretty neat system for a free-ranging dolphin society in general. If the mother was lost, a calf could at least recognize other members in the group as closely related, and they would recognize the calf. The complexity of this dolphin society never ceased to amaze me.

Delving into the mind of another species, especially a nonterrestrial, nonprimate one, is challenging. On our research boat we often play charades and I must admit I not only love the game but I am quite good at it. I love to observe how different minds communicate information. There is the player that tries to play by communicating the whole concept, such as *Jurassic Park,* by mimicking a dinosaur. Then there is the person who does the second word first, by walking through a park, then breaks up Jurassic by syllables, Ju-ras-sic, and rhymes every syllable, or does "sounds-like" rhymes as cues. What is particularly interesting is the person who does one thing, like "dinosaur," and then repeats and repeats the signal in only one way, and cannot fathom why no one understands such an obvious signal. Simply, their technique lacks information and clarification. The player is expecting that everyone thinks like him and therefore the communication is simple and clear, and this person usually fails in the attempt. They are not paying attention to the audience signals, encouraging them with eyes and hands when they get close or on the right track. Charades is an example of DST in some ways as you create the meaning together. Good charades players adjust their signals to meet the needs and produce the desired results. Try it sometime, and you might learn new things about how people attempt to communicate.

Individuality and knowing the history of, and your relationship to, the other entity certainly weighs on the success of a communication or the failure, as we all know from our personal interactions with the world. In science we use statistics to measure things, usually to find the "average" or norm of something. However, in behavioral analysis it is often the individual, or outside-the-mean activity, that is interesting and pertinent to our question. When we "average" out patterns for complex animals, as we often do in statistics, our ability to understand the diversity of individuals and roles in an animal's society is reduced. It is the analysis of a significant event—such as the death of an alpha chimpanzee male and the shift in dominance or the death of a matriarch female elephant with decades of experience—that may mark a significant change in a family or society. Behavior, by its very nature, is dependent, not independent. In our own human cultures nonconformity and improvisation, in the face of challenges, are often more telling of our culture than a set of norms. David Lusseau and colleagues have described the importance of individual dolphins in a society where networkers, brokers, and individuals who socialize and potentially encourage interactions and exchanges of information provide a critical networking role between dolphin communities. When these individuals are lost, the community suffers and becomes more isolated. As for other social mammals, in a dolphin society an individual matters.

One of the results of spending so much time underwater watching the dolphins over the years is that it has allowed me to interpret their behavior more accurately from the surface. In most places in the world surface behavior is the only information available to a researcher. In the Bahamas we work in the water most of the time so we don't have to assume or speculate on what is going on, we actually see it underwater. Of course there are murky days, and days when the dolphins are on the move when we slip in and out of the water, sometimes over and over, only to find that they just keep traveling past us. One day we watched in murky water as a group of male spotted dolphins chased a bottlenose dolphin displaying open mouths, head-to-heads, and the coordinated squawks and jaw claps. We couldn't keep up with them in the water or see them very well so we got out of the water and followed them in the boat. From the surface we saw aerial side breaches and close-up penis displays of bottlenose dolphins while the spotted dolphins were lifted halfway out of the water by the bottlenose dolphins' energetic behavior. There were side rolls and the ventral sides of dolphin bodies at the surface (which reminded me of the avoidance behavior of female grey whales during mounting attempts by males). The young mottled males were the ones getting most of the sexual attention from the bottlenose dolphins as they displayed bubble blows, chuffs, and tail slaps, all typical behaviors that occur during aggression and that we had seen before, but underwater. It was nice to be able to translate at the surface as well as underwater.

On August 20 we stopped to swim at a reef on the south side of the island and we were joined by a large manta ray that allowed us to swim with him for hours. Patches of squid eggs, in their circular pattern, were attached to the sandy bottom along the deepwater edge. Tropical Storm Dean headed toward us as we finished with our field season early on August 23. At the end of the season I once again tallied the dolphins we had seen and reluctantly I acknowledged yet another devastating loss, Rosepetal was missing this summer and not with her well-known friends. The entire Rosemole family was now gone and like my own family memories, faces and antics of the Rose family flashed back in my head as we headed back to Florida, grateful for another field season, but sharply mourning our losses.

To break from the norm I had a special fall planned. I was heading to Greece to join colleagues in the Gulf of Corinth to observe common dolphins, striped dolphins, and Risso's dolphins that intermingled. As we drifted in the Gulf of Corinth, under the shadow of Delphi, I was reminded of how many millennia humans have interacted with dolphins.

As we left my colleagues to their studies I headed to Delphi to play tourist. What was it like in 3000 BC in Greece? Did the local dolphins lead Greek sailors into the harbors below this mountain that loomed above us and harbored Apollo's sanctuary and the Delphi oracle, the ancient version of the Internet? September 11 dawned bright and clear. After a stroll down to the ruins I walked back up the steep hill with a friend and took note of the flags that waved in the wind as a global symbol that all people are citizens of the world at Delphi. I reveled in the harmony in the world, and the location I was in, taking in our ancient human history. As we walked by the shops in town, a shopkeeper waved us in to see the news on his small and dusty TV. The World Trade Center had fallen and the world had changed forever.

2002–03—Winter Work in the Bahamas

Mid-May brought a late spring cold front, making the wind and water cool. Fourteen-foot swells in the Gulf Stream battered us as we crossed to the Bahamas. Mid-June brought no better weather, with tropical wave after tropical wave, hiding the summer solstice sun for ten days. It was a sign of things to come in the form of tropical waves and storms all summer.

Despite the high seas, early spring was filled with interesting observations of fish-hunting tactics. Although the dolphins routinely terrorized the large horse-eyed jacks on the dolphin wreck for target practice, now they focused on the smaller bar jacks, a more reasonably sized prey. I watched Trimy and her calf Tyler, and Snow and her calf Sunami, as they darted about after fish. Liney and another adult male were herding Snow in between her bouts of chasing and disciplining her calf on the bottom.

A few days later we saw another group of dolphins chasing a fish ball. Geo, Bishu, Heaven, Jango, and Little Hali darted about separating a bar jack from the school and taking turns chasing it. One loud clap, possibly a hit, was heard during the chase, although it looked like practice today. Bishu got serious and chased one under the boat and then brought it by as we watched her swallow it headfirst. Sometimes the bar jacks took shelter with the schooling horse-eyed jacks and the dolphins stationed themselves on the periphery, scanning about and trying to find them.

We knew the dolphins were adept at dislodging fish from the side of the boat, from behind our cameras, and away from our bodies if needed. One lazy day we were anchored on the dolphin wreck when in came a group of juveniles. A small ballyhoo was sheltering against our cook and the dolphins buzzed and blew bubbles to dislodge the small morsel from her body. It's lucky we knew their intent because it might have appeared as if they were aggressively challenging the human, but after years of fish swimming behind our cameras and up our swimming suits, we knew what they were after. Occasionally the dolphins hunted a shrimp in the sand to eat (a rare occurrence) or terrorized it by holding it in their mouths and then releasing it for more harassment, as they did with their occasional filefish victims. Sometimes the shrimp, like a fish, sought shelter among humans or cameras, and eventually escaped, destined to be someone else's dinner.

June, often a turbulent month, drifted by with its usual squalls and waterspouts. Large storms to the north sent monster swells reeling across the ocean toward the northern Bahamas. Although we anchored comfortably most of the time in the swells, it was impossible to swim adequately during our jumps in the water to follow dolphin behavior. Dolphins frolicked around our boat while we bobbed in the large swells, hoping to get at least an identification photo as they swam by. We were also hoping that we could get out of the water and back on the boat's dive platform without getting pulverized.

After fifteen years we thought we had seen all stages of pregnancy, but now we had a new insight, this time provided by an unhappy bottlenose dolphin. Natasha, a female bottlenose dolphin, was pregnant. In fact she was very pregnant. Although we had seen various stages of pregnancy every summer, Natasha looked like she was ready to pop. How far could a dolphin be stretched? She was gigantic to say the least. Natasha was very curious at first, but later started herding us back to our boat with the assistance of one of the juvenile male spotted dolphins, who was also posturing, although he seemed too young to be so aggressive. Natasha whistled and blew bubbles as she swam with us and then displayed an open mouth as she herded us back to the boat. We take these as serious signals from any dolphin and got out of the water. And this day Natasha seemed in a very bad mood. Finally Nightmare was born, and Natasha returned to her normal trim self once again.

Pregnant spotted dolphins can also engage in aggressive behavior when needed. I watched anxiously as Little Gash was being herded by a group of males; her intense tail-slaps to their bodies were the kind that made me wince in the water. Little Gash, even though she looked about seven months pregnant, was head-to-head with Poindexter. Another adult female, Hook, was chased and held down by Slice and his coalition on the bottom. I remembered encounters I had seen when males had open wounds and scrapes. Apparently when they fight, they fight hard.

July rapidly went by as the months often do; mornings start early and the days end late. Stretches of good weather gave us good data collection opportunities, but made for exhausting and sunburned days. I was anxious for August to arrive since Marc Lammers was coming out again with not only Spin and Spot, the high-frequency units, but with a towed hydrophone array that he had used with the spinner dolphins in Hawaii. As we towed a line of hydrophones behind our moving vessel, we zigzagged in and out of the traveling spotted groups. This technique had worked well for Marc in Hawaii, but spotted dolphins are notoriously quiet while traveling and the sound sampling was sparse. Later that August we entered the water with both Spin and Spot, hoping to coordinate our sampling angles in the water, which is more difficult than you think. Sometimes we got in the water with the heavy equipment only to find that, although there were loud and frequent vocalizations, the visibility was murky and video was useless. Other times we got in with crater feeding bottlenose dolphins and got great samples, head-on, as they approached our equipment. Distracted by my sound sampling, a ten-foot bull shark suddenly appeared next to the group of dolphins. Not wanting to repeat my bull shark experience of 1986, even though I had a hefty piece of equipment to block a charge, I opted out of the water. Once, on a night drift, I had observed a large tiger shark under my feet as we all struggled to get out of the water. I was not about to drop my negatively buoyant equipment in one thousand feet of water that night, and today I was going to calmly get it out of the water and away from the already too close bull shark.

The first day of September we came upon two bottlenose dolphins near the deepwater edge. Hoping to get a chance to get some head-on shots of diving dolphins (by getting on the bottom before them, and pointing the unit upward), we jumped in. The bottlenoses sometimes hover stationary in the water column and rotate their heads in a circle. This is one of those "quiet" times, where I suspect they might be using their ultrasound (or just listening). After free diving a dozen times to set myself up to capture a head-on signal, I was finally rewarded by catching a short segment of a bottlenose dolphin rotating her head in a circle. But I would need many more samples to determine what, if anything, they were doing with their sound.

2002 drifted into 2003. Where does the time go? Where do the years go? My life had a routine seasonal pace. May through September was the fieldwork. I still was doing every trip, about ninety days every summer, hurricanes willing. In the fall I would take a break and go hike in the woods somewhere. Winter we analyzed data, updated the identification catalog with photos, reviewed video, wrote grants, and then started getting ready again for fieldwork in May.

When people come out on our research boat, all they see is the field part of the work, which is sometimes fun, but 95 percent of the time it's just hard work. We are working until sunset and then to midnight reviewing our videotapes, night after night. I would often go home during our long weekend turnarounds and sleep for a day. It's the kind of physical exhaustion that makes you feel drugged; you couldn't get up if you tried. Month after month, year after year, this was my routine. We jokingly tried to "create" another month in the year on our office wall calendar. I named it "Extru-ary" and placed it between January and February, but it never really worked.

We also started doing winter trips to Bimini, on Great Bahama Bank south of our normal study site. Since we had a boat, and knew there were dolphins down there, we decided to try to do some winter work, as our normal study site is inaccessible in the winter months, due to predominant northeast winds. Could the dolphins in our regular area be seen in Bimini? Suppose they intermingled, shared their genes?

I remember the first time I got in the water in Bimini. It was actually late one summer, and although I didn't like leaving our field site in the summer months, a hurricane up north was causing unworkable swells for us, so we thought Bimini might be a place to explore, given its more protected location. I slipped in the water with my normal video gear. As a mother and calf approached I looked at the adult and estimated she was probably around twenty years old and the calf two years old based on his size and lack of spots. I felt shocked; I didn't know who they were. Who were their friends? How many calves had this adult female had over her lifetime? I was so used to seeing individuals I knew, had grown up with. I knew Little Gash's friends, her calves, I knew Stubby's coalition members, and I knew their personalities up north. But I didn't know this mother's history! It was both funny and startling. I found myself in the water asking this mother, "Who are you and what has your life been like?"

Working in Bimini in the winter had other pleasures besides the dolphins. The winter winds were cool, and the air was not torturously humid as it is in summer. At night we anchored our boat in the lee of the island, opened the hatches, and slept. Working days were shorter in the winter. In the summer we were used to working in the water until 8:00 or 9:00 P.M. Here in the low light of winter the sun rose late and set early, leaving us evening hours to fill with activities. We kayaked to the beach; went into town to the historic Angler, the famous Ernest Hemingway bar that has since burned down in a fire, priceless photos and all. When we anchored up at the northwest end of the sandbank we swam over to the razor-sharp coral island and climbed around the lighthouse that provided a beacon to oceangoing travelers. Although Bimini was interesting for winter, I had no desire to split my summer field time between the two sites. I was committed to my long-term focus on the community I knew up north, which had a richness of details that only time could yield. We eventually gave up our winter work and left our photo-identification information in the hands of newly located researchers there.

But the question of how much exchange there was between study areas and who was siring offspring remained. For a long time we had no way of identifying which males were siring offspring in our study area. We could tell who the mothers were by observing nursing, close swimming, and other surface observations (later confirmed by genetics). But dolphins are notoriously promiscuous and finding out who the fathers were was like looking for the proverbial needle in a haystack. Fecal analysis was a technique known in the primate world, and Kim Parsons had started fecal analysis with bottlenose dolphin samples from our area. Michelle Green, an eager young student interested in mating systems, stepped up to the challenge of trying the technique with the spotted dolphins. We continued our in-water protocol to collect fecal samples for DNA as we swam along, matching them with our known individuals. Over a period of eight years we collected, labeled, and processed our many fecal samples. Combined with our long-term life history database we soon had a list of potential fathers and the oldest males, Romeo and Big Gash, were winning the race. For years I had watched young mottled males competing for access to females and fighting among themselves. I had often wondered if this sexual behavior was simply for pleasure, for establishing rank or displaying to females, or for real breeding. We knew from other studies with pantropical spotted dolphins (from samples of dead dolphins in tuna nets) that males had to be "fused," or at least fifteen years or so, to breed. So it looked like all that fighting among the younger adult males is either practice or display. The older male strategy of "hanging out" with the females and helping raise their young might be a viable reproduction strategy in the dolphin's world.

This summer went by at the normal pace. Spring storms made our May and June rougher than normal. By July we were into our routine, mine in the water with UDDAS sampling head-on echolocation of bottlenose dolphins and signature whistles of spotted dolphins. We collected more fecal samples, more identification shots. I had a new student working on the bottlenose dolphins, Cindy Rogers, and she eagerly familiarized herself with the intimate identification marks of both species. The bottlenose dolphins were finally getting comfortable with us in the water, probably because of the new generation of calves and their positive experiences with humans, so occasionally we saw bottlenose social behavior, which had in the past been rare. Some of them were actually getting friendly and playing seaweed games with us. It only took nineteen years, but I wasn't complaining. I had always wanted to get more samples of bottlenose-only behavior, to be able to use for a comparison of their behavior when they were with the spotted dolphins. I had always suspected, because of their tattered dorsal fins and raked bodies, that bottlenose dolphins were of a different, and more aggressive, temperament than spotted dolphins. And in the mid-1990s I had seen the extreme aggressions of a male group of bottlenose dolphins close-up. But most of our observations of the bottlenose dolphins underwater were in the presence of spotted dolphins when they foraged, traveled, or negotiated conflicts. Many of the "friendlier" bottlenose dolphins had spots on their underside, which meant possible hybrids, so we continued collecting fecal samples hoping to get some samples from these individuals.

Although females are vulnerable during late pregnancy, and often seen in the company of groups of male escorts, they can both challenge and fend off large sharks. After watching three very pregnant females, Uno, Priscilla, and Caroh, ride the bow, we decided to enter the water to take a better look. As I slipped my fins on near the stern, I peered around the corner looking forward and saw something out of *Jaws* as a giant thrashing shark tail slashed the water. The pregnant females had left the bow and gone after the shark, apparently annoyed by his presence. When we finally got in the water, Mugsy (about eight months pregnant) and Zot (around seven months pregnant) were lazily playing on the bottom.

Even the nurse sharks occasionally surprised us with interesting behavior. One July day while observing Stubby, Horseshoe, and KP in the water, a dark figure appeared at my side. The now visible nurse shark began to eat, or at least skim through, the defecation of the dolphins. During other moments the shark was catching small snappers and grunts. Who knows, maybe he saw us focusing on and collecting their defecation and sensed a potential meal?

Although watching underwater behavior seemed routine at times, there was always something that happened during the summer to fill in a missing piece of information. We saw more and more examples of who was in charge in the dolphins' group. Leadership in a dolphin group is based on age and experience. A large group of old males often escort pregnant females or nursery groups, presumably for protection. The most experienced female leads mother-calf dolphin groups without males present. The eldest juvenile who usually has the babysitting role leads juvenile groups. Age and experience seem more of an asset than a liability in wild dolphin society.

Occasionally we found the dolphins in little-explored areas of the sandbank. This summer we were far up north on Matanilla Reef, a pristine reef on the windward, northern edge of the sandbank. It was early July and the weather was conducive to moving around more, so we did. Much to my surprise there were some of our "southern" group way up north. What in the world were they doing here? Could this be why we don't see them as much in the central area? Is their habitat a longer stretch of north to south rather than east to west as we see with our other dolphins? We had hints of this possibility years ago when Triangle, a southern dolphin, was seen far north, but today it was Flare, Roxy, and many of the southern gals. They bow rode, played sargassum games with us, and led us into the reef itself while our research vessel hung back, unable to follow us into the shallows. I wondered if the dolphins were surprised at seeing us here outside our normal area of contact?

We monitored the new calves throughout the year, noted which females were looking pregnant, videotaped examples of juvenile behavior, adult behavior, and everything in between. We refined our collection of the high-frequency information, gathering more and more data and behavioral contexts for vocalizations. Sometimes we got caught in the water when lightning began and I can tell you it's a frightening feeling with a piece of metal in your hands. We waited for the movement of the tides, the moon, or the weather for an opportunity to travel to our normal areas. We trolled on the edge for our supper when time allowed. By early September Hurricane Fabian was heading our way with 140-mile-per-hour winds, so we ended the season and left Fabian and multiple tropical depressions in our wake.

Our last night we watched the electric summer lightning storms form over Florida and our homes far across the Gulf Stream. With our night-scope we peered into the Andromeda galaxy, visible to the naked eye as a fuzzy ball, but thoroughly clarified by the use of starlight through a scope. I remember when the vast majority of the public thought it was crazy to conceive of life-forms on other planets. As I looked up in the Milky Way, the densest part of our galaxy, I remembered seeing shots from the Hubble Telescope of galaxy after galaxy in our known universe. I wondered how there could *not* be life out there; it seemed impossible. We know now that there are many star systems with planets, some with water and ice that potentially house the type of life-form we envision to be possible. I reflected on the diversity of life-forms on this planet and I thought how amazing they are and how both similar and different they can be. Then I realized that I am working with one of the most intelligent forms of life that we know of on our planet—dolphins—and felt very small and humble to be there.

2004—Two Decades Later

As I started my field season in 2004 I realized it was my twenty-year mark. I had managed to be out in the field every year, not missing the changes in spot patterns of individuals or the continuity of their behavior. I had incorporated some cutting-edge technology into the fieldwork that was illuminating some of the larger mysteries of dolphin life. And as well-known dolphin researcher Dr. Renee Busnell remarked during a lecture I presented in France, I had (with much help and luck) found the funding to keep a field project going for twenty years, a feat he thought as remarkable as the work itself.

June and July went by with the normal increase in heat and waterspout activity. The end of July brought storms so fierce we found ourselves running from unusually large and dangerous waterspouts. Early August found us tied up at the dock a few nights due to more severe storms. The tropics were heating up with Tropical Storm Bonnie and Tropical Storm Charley forming, and there was a hurricane watch for the Florida Keys. Concomitantly we found ourselves relaying messages to the Coast Guard as boat after boat up in our remote area got in trouble.

As the field season flew by I was concentrating on my efforts on the high-frequency equipment to obtain more samples of sound and behavior. My femme fecales, the endearing term we used for my female graduate students, were becoming proficient at collecting dolphin fecal material for genetic analysis. Occasionally they could even talk the dolphins into giving them samples, or so they said as they followed an unyielding group of adult males. Since we really needed their samples for paternity tests it was a good thing. Like their signature whistles, males do not easily yield up their defecations. When they do defecate, they are often in a jumbled group of males, making it unclear whose fecal matter is whose and defeating the purpose of collection. But our long-term fecal collections were paying off. Michelle had already discovered that males had to be fused, or fully adult (at least fifteen years and more likely eighteen years of age), to sire offspring. Romeo and Big Gash were both fathers two times over. Romeo was even grandfather to Tristan, Tyler's first calf. Navel, who was a calf in 1986, was also a father. This meant that Poindexter, Geo, and Latitude, of similar age, could also be siring offspring, but without genetics it was difficult to confirm. This also meant that Luna, Gemini, and Pair One had potential offspring in the gene pool. Extended families are the norm in this society and the connections and networks were likely more dense than we could even track. We had now verified, through genetics and association patterns, that this society was divided into three groups: the south, central, and north. Although males mated mostly within their own social units, it was looking like individuals from the north and central group crossbred sometimes, as did the central and the southern group. We now knew from the variety of fathers of Paint's offspring and other females in the group that the dolphins had a promiscuous mating system. Females tended to stay within their own social units, while the males dispersed throughout the larger range, a very typical mammalian pattern. In addition we had seen, over the years, immigration of new juvenile males into the groups, probably from other communities in adjacent areas.

I thought about the individuals who were still around that I met in 1985. Rosemole and her offspring were gone, but Little Gash and Mugsy, her good friends, were around and now mothers multiple times. The Gemini family was still here, boasting a family of four offspring. White Patches was long gone as well as others in her circle of friends. Romeo and Big Gash, lifetime coalition friends and probably the oldest dolphins out here, were still around. Horseshoe and White Spot still frolicked in the waves, as did Punchy and Big Wave. I thought of the new lives I had seen come into the group: Little Hali, KP, and Caroh, our star ambassador for the two-way work. New characters and personalities were injected every year, but some were lost. Poindexter and KP, now older males, were taking the role of instructor for the new male calves and juveniles in the group. Laguna was now old enough to babysit since her mom Little Gash was pregnant again. In fact, all the females in our group who were sexually mature and could be pregnant were pregnant! Dolphin life, it seemed, had its own rhythm. Age classes had their own responsibilities. Mothers had their teaching agendas. Males continued to battle for females and rank in their groups. Everything was just fine.

By late August hurricane season was really ramping up and we found ourselves ending the field season early. There were storms lining up to our south and to our east, all of them pointed toward south Florida. Excited and grateful as I was, I found myself tired mentally and physically. The many hats I wore were draining my energy. I was still frustrated at my inability to get some new two-way communication technology in hand. Although four years of pilot work in the field had shown me how it could work, I was frustrated dealing with engineers, software writers, and colleagues who promised help, got paid for their time, but in the end could not produce a viable product to use in the water. In hindsight it's like any project. You really have to have enough resources to do it right, and we didn't. Various benefactors and foundations had helped us start the work, covered our field time in the early years, and cheered me on, but in the end it wasn't enough. Faced with my own unlikely abilities to do another twenty years of fieldwork, I was determined to still build a two-way system as soon as possible. But Mother Nature had something else planned for me. Little did I know that storms were brewing that would upset my plans.

~ 9 ~

Three Generations:

Grandmothers and Grandfathers

There is no moral precept that does not have something inconvenient about it.

—Denis Diderot, L'Encyclopédie

2004–06—Hurricane Hell: Frances, Jeanne, and Wilma

In early September of 2004 a tropical system was brewing out in the Atlantic and she was named Frances. Like Hurricane Andrew in 1992, this storm was on a track directly toward us, and it didn't seem to be changing course. I had a bad feeling about this one. Sure enough, the weekend after we had unloaded our boat we were frantically evacuating to escape Hurricane Frances. "This one is coming right for us," I told my board of directors as I notified them in an e-mail that we were securing home, office, and boat. The last few years we had been ending our field-work in early September, partly because peak hurricane time is the first two weeks in September, and partly because after eighty or ninety days of fieldwork we usually had obtained a reasonable amount of data.

When preparing for a hurricane you know you have to pack up, evacuate, have your supplies on hand, and empty out your refrigerator because of the inevitable power loss. It's a stressful event, but organized stress. This storm was different. As I packed my bags and loaded my car with my cats, tax records, and my guitar, I remember thinking I could lose everything—my condo, my office, and our boat. As I drove out of my driveway and off the island I had visions of my concrete-block house as rubble. It was a scary feeling that made me think of the many people who have suffered from natural disasters and the multilevel traumas that occur including the loss of stability, normal patterns, and sometimes the loss of loved ones.

Since I live across the street from the beach I had to evacuate, but friends had offered protection at their inland home during the hurricane. As we waited for Frances to approach we barbecued the sure-to-spoil food in their refrigerator. Champagne and French chocolate came out for my "Happy Birthday" in the now powerless and sweltering house. As Frances reached her full fury, we eagerly brought out a bottle of tequila and sipped while we listened to the winds rattle the shutters and outside doors. I wondered what the inside of my home sounded like? Would it be there when I returned?

Hurricane Frances hovered and then she went stationary, as sometimes hurricanes do, for twelve hours. During the worst of the storm we went into our closets, literally. In the spare bedroom my two cats, who readily sensed that their cat carriers were open for a good reason, hopped in. As the barometric pressure dropped and the storm's fury hit I anxiously listened in the dark to the radio broadcasting the timing of the eye wall, the strongest part of the storm. Now is the time to go in your "safe room" they would say, but we were already in our safe closet.

One end of the hurricane eye had hovered over Palm Beach County and the house I was in while the other end of the sixty-mile-wide eye hovered over our study site and the dolphins. After regrouping three weeks later after power had been restored, I decided to take my hurricane shutters down. We unpacked the office including every book and piece of paper secured in plastic bags. Captain Pete had just brought the boat back from the middle of Florida, fighting his way down the Intracoastal

Waterway in between sunken boats. But another even more powerful hurricane, Jeanne, was hovering offshore and the weather station had predicted that Jeanne would move north to northwest toward the Carolinas and miss Florida. But never underestimate what a hurricane will do, especially if it is hovering just east of you in the Gulf Stream. Jeanne revved up her engines and headed directly toward us, taking the same path as Frances. We packed up our homes, office, and the boat once again and took shelter. The eye struck within two miles of the eye of Frances, following her path of destruction.

Palm Beach County had taken a hit, with trees and small structures down, but it could have been much worse. Through the winter we heard about the Bahamas: how the storm had broken through the jetty at West End; how lobsters, either thrown or left on a waterless reef, were wandering around on land. I wondered about the dolphins, how had they dealt with it? Did they leave the area or get exhausted fighting the waves while trying to breathe for twelve hours in the mist-filled air? I wouldn't find out until next May! How ironic. It was the twenty-year mark for my fieldwork with the dolphins. I had picked twenty years as the minimum time to observe the society and hopefully a third generation of individuals, and I wanted to see the work continue and prayed that the storm did not end that possibility.

The 2005 field season was finally here and I was anxious to get out to the Bahamas to see how the dolphins had fared during the hurricanes. It was already May 31, but we found ourselves still stuck in the boatyard due to work delays, anxious but unable to leave for our first trip. Finally after a late-afternoon departure we were out of the boatyard and on our way to the Bahamas. Summer was already in full swing; tropical storms were already brewing. Although we looked and looked all summer, there were many dolphins missing. We looked north, we looked south, we looked farther east than before, but found none of the missing dolphins. I remembered the previous summer and how all the female dolphins that could be pregnant were pregnant. I wondered if they somehow sensed the impending 2004 hurricane season. By mid-July Hurricane

Emily, with 135 mile per hour winds, was heading our way. After last year it was hard to imagine another severe hurricane season, but it looked like we were heading for one.

By August 12, Tropical Storm Irene was approaching, followed by Katrina, which formed over the Bahamas on August 23. As we left our anchorage to head for Florida and await the passing of Hurricane Katrina, we measured the water at ninety-four degrees Fahrenheit. Unheard of, what did it mean? Katrina took a minimum swipe at Florida and then popped out into the Gulf of Mexico, where it devastated New Orleans in the weeks to come. But we were monitoring our own devastation in the Bahamas as we counted all the dolphins that were missing.

When I tallied our missing hurricane dolphins it was heartbreaking. The list was long and it spanned fairly equally across gender and age class. Many dolphins I had known for twenty years were gone. Except for Sunami, the entire Snowflake family from the north was gone. Gemini's whole family, except Geo, from the central area was gone. Not only Gemini and her girls, but also Heaven, Punchy, and Big Wave were gone. The list went on and on. I jumped in the water with my mask on and cried. I could taste the salt from my tears like the dolphins taste seawater. But I knew this salt was not coming from the ocean, it was seeping from my heart.

I couldn't stand the pain of thinking that I would never see them again, never play with Galaxy again or hear little Galileo's excitement vocalizations as she swam in circles around me. I remembered last summer in late June, when a group had come to bow ride. Gemini, Galaxy, and Galileo were all together. Gemini swam inverted after a rambunctious Galileo, while Galaxy, her sister, stuck close by. Galaxy gently nudged Galileo with a friendly beak-to-genital motion, while mom was stretching and tail-slapping on the bottom as Heaven frolicked nearby. Now they were gone, likely forever.

I imagine many of them were just exhausted and drowned after battling waves and wind for twelve hours. At greatest risk were the young calves struggling without experience and pregnant mothers with high energy needs. They were probably lost either by the direct impact of hurricane winds or by a secondary effect such as losing their food source over the past winter. Where do resident dolphins go with a 500-mile-wide hurricane approaching? Do the elders have long-term histories and memories of places to go, like the elephant matriarchs that find watering holes in years of extreme drought? Do the dolphins have such a plan? Do they even have options? The impact was incredible. More than 30 percent of the spotted dolphins and known bottlenose dolphins were gone. How could such a thing happen to a healthy, free-ranging community of dolphins? I couldn't believe it.

It's possible they were displaced and just moved elsewhere to find more food. Decades ago, in the Channel Islands of California, there was a displacement of pilot whales in the area, due to El Niño and the change in their food source. Deep in my heart I felt that many of my dolphin friends were gone, dead from an extreme hurricane season. It was hard to imagine that a natural disaster could wreak such havoc on a healthy natural population. I was reminded, once again, why it is so important to not let any species, or any community or ecosystem, get close to the brink of extinction due to human impact. Nature already provides enough of a challenge for animals in the wild.

These events reminded me not to take things for granted. When we lose a loved one we recall that every day is precious and the time we spend with loved ones may not last. In one last bid for the planet, in our current crisis of human destruction, we are advised to not take the Earth and its health for granted.

There weren't many great discoveries in 2005. Most of our time was spent looking for the dolphins. When we found the dolphins they were traveling, always traveling. Their social structure was a mess. Many major groups had lost at least one member. The northern group, already small before the hurricane, was reduced to half its size—around twelve animals. In the past, when we saw the northern group, grandmother Snowflake, her daughter Snow, and grandson Sunami were there. But now as we saw Paint and Brush and their surviving offspring, I sadly looked around for Snowflake and her clan, but they were missing. Three generations neatly bound by the same forces you would find in any human community ripped apart.

If 2005 had been my first summer of research I never would have believed I could observe these dolphins for twenty years given the way they acted this year. We saw very little behavior that summer. The dolphins' social structure was in tatters. Frankly, I didn't really care about recording their behavior that summer; I just wanted to find them. By the end of the summer we had found no missing dolphins. Another active year in the Atlantic, and we had dodged hurricanes all summer—so many that the rarely used Greek alphabet was used for the next set of names for storms. It was a record year.

In early September we went home and mourned our losses from last year's hurricanes and the devastation of our dolphin families. But that was not to be the end of it. A freak October hurricane, Wilma, was raging through the Caribbean, and after it wiped out Cancún and the surrounding areas in Mexico, it set its sights on south Florida. Not again! But this time the storm was coming in the back door, from the west. Because of the direction of the storm my area was not under evacuation so I stayed home, but being a weathered sailor I put up my hurricane shutters again and Captain Pete secured the boat locally. As I sat huddled in my bathroom with my power out and my flashlight dying, I listened to my shutters ripping off my front door and I felt like I was in a tin can and being opened for dinner. Whoops, the weather service had miscalculated, the storm was getting stronger as it charged from west Florida to the East Coast. Hurricanes gain strength over water, and the forecasters had missed that, with one side of the eye wall over the Atlantic and the other side over the middle of Lake Okeechobee. Wilma had grown. So fierce was it that the southwest side of the eye wall, usually the weaker, was the stronger, so I braced for the other side as the first side of the eye wall passed. I did go outside during the eye, and along with my neighbors picked up pieces of debris that were destined to become flying projectiles. Then I took shelter again in my bathroom. As the last of my flashlight's beam dimmed and the sweat rolled down my back, I called a friend with my fading cell phone swearing that I would never stay home for a hurricane again. Temporarily worried about my own survival, I forgot about the potential impact of Wilma on our study site. Although the immediate focus was on south Florida on the news, Wilma had continued to the east toward the Bahamas and once again scoured our study site and the dolphins' home.

We returned to the ravaged study site in May of 2006. The good news was that there were no new missing dolphins. It was my guess that the stationary posture of the first hurricane, Frances, was the main culprit of our dolphin mortality in 2004. Once again we looked and looked for our lost dolphins with no luck. But we had seen an influx of new bottlenose individuals last year that Cindy, for her dissertation, was monitoring. Where did they come from? How would they integrate? It would take years to find out.

People often ask me what the dolphins do in a hurricane. Of course I don't know because we fear for our own safety and head to land. In a minor hurricane or tropical storm I imagine they surf. But with the sandbank only fifteen feet deep in many places, the swells of a large hurricane make the sandbank an exposed and dangerous place for even the smartest of sea creatures.

So nature took its course, and males, females, and all age classes of the dolphins were still missing, including whole families, well-known individual units of male coalitions, and members of a female group. Although I felt sad about the shake-up in the dolphin community, I couldn't help feeling extremely grateful for the previous twenty years of stable observations. Establishing a relationship as I did in the early years probably would not be possible now because of the dolphins' changing priorities posthurricane. I always knew how lucky I had been to find a place in the open ocean where I could observe dolphins on a regular basis. But until these drastic changes occurred, I had not fully comprehended what a perhaps unrepeatable privilege it had been.

I wondered why Geo, the lone male offspring, survived and the rest of Gemini's family didn't. Was it because he was a male and hanging out with his male coalition? Geo had always been a bit of a peripheral dolphin at the low end of the totem pole with his juvenile male friends. Burdened by an annoying remora, which clung to his disease-ridden skin, Geo was a bit of an outcast for a time, but gradually integrated with his male coalition, which included Mohawk and Latitude. Sunami was also the sole survivor of the Snow family. Did the males have different strategies for survival? It was sure looking like it.

The weather was turbulent all summer, but at least there was little to no hurricane activity in the Atlantic, a welcome break from the last two devastating years of storms. In mid-July we found ourselves observing sharks instead of dolphins, as the now well-trained and hungry lemon sharks greeted us each morning, circling our platform expecting the chum and fish they would receive from the local shark feeding boats. Boy, things were topsy-turvy—trained sharks, irregular dolphin behavior, and no hurricanes.

2007–08—Living on the Edge

On the first day of the 2007 field season the Gulf Stream seas were so high that we completely missed our customs appointment in West End and anchored off the island that night, recouping from our physical abuse. It wasn't even until early June that we were able to travel up north to the dolphins' wreck due to the continued high seas.

We continued to look for the dolphins all spring, but no missing spotted dolphins were found. A new master student of mine, Tiffany Adams, had taken on the project of comparing identification marks to photographs taken by my colleague Diane Claridge in an adjacent area in the Bahamas, the Abacos. If we were to find one match, then everything would become possible. But it was likely that these individuals were gone; gone for good. Why were new bottlenose dolphins able to move in to the area? Over the years we had documented, with Kim Parsons, the exchange of bottlenose dolphins between our field site and the extreme east and south sides and Little Bahama Bank, making the bottlenose dolphins' range larger than the spotted dolphins'. Late August brought Hurricane Dean our way, so we ended our season early due to yet another storm threat moving through our area.

Although new bottlenose dolphins had moved into the area immediately after the hurricanes, it wasn't until 2008 that we were finally able to verify that some new spotted dolphins had arrived and were trying to integrate with our regular group. Dr. Michelle Green—fondly known as Dr. Poo on board because of her groundbreaking work extracting DNA from the dolphin's defecation—and I were in the water during midsummer and noticed a group of four spotted dolphins we didn't recognize. They appeared tight, like a male coalition. Michelle and I were baffled both in the water and that night reviewing video. Who are these guys? They had great marks on their head, large spots on their body—surely we must know them? Feeling a bit insecure, as if had we lost our "touch," we asked Cindy, our photo and database manager, to take a look. If she knew them we would be very embarrassed. But much to our relief, on a couple of levels, they were new dolphins. Through the summer we saw a mother and calf who were also new to the area. So things were looking up, the spotted dolphin niche was being filled, both with new calves and with immigrants. But where did they come from? How did they find out there was space? Did they get displaced by the hurricane as well and were in search of new digs? If so, might the dolphins we know still be out there somewhere, looking for a new habitat? It's possible that there wasn't sufficient food for everyone. The impact from habitat destruction and subsequent changes in the food chain can drive resident marine mammal communities to new areas. We have, after all, had dolphins return after ten years to our area. So we continued to look and hope that somewhere out there Gemini and her young family had found a new home, perhaps with Snow's group.

Posthurricane years also showed the complete cessation of interspecies interaction between the two dolphin species. Where there was once a 15 percent level of diverse and complex interaction between spotted dolphins and bottlenose dolphins, in the summer of 2007 we saw virtually none. The summer before, 2006, we saw only a small amount. There was no doubt that things had shifted. As in most natural situations the dolphin community regained some stability and normalcy after a few years, but the dolphin's social structure was still abnormal, which made our work very frustrating and difficult.

By 2008 we had seen a glimpse of normal behavior again from the spotted dolphins and by the end of this season they were acting as they had in the past. It had taken the dolphins more than three years to recuperate and begin to act and reproduce normally, with ten new calves born, well above the yearly average of four. I celebrated twenty-five years at sea, and as I watched the now normal behavior patterns of the spotted dolphins stabilize, I was relieved. I realized that the two-way work was now again possible. A healthy crop of young juvenile females was available and exuded some of the same personalities Caroh and her friends did in 1997. Caroh, now a mother herself with Copper, still swam over to us in the water and gave us a nod of recognition. She even looked longingly at a scarf in the water. But her life had changed, as it should. After multiple years of pregnancy Caroh was finally a successful mom. Like another dolphin I'm recalling, White Patches, Caroh was a tomboy, but she finally managed to be a great mother. Who knows, as we restart the two-way work perhaps Copper has it in him to break new barriers and be a young male interested in the system; after all it's probably in his genes.

I wondered if these remarkable abilities that emerged during the two-way work came from individual variations and personalities or from their large brains and type of intelligence. In a general sense, intelligence is the ability to survive in one's environment. By that measure, most species are intelligent. However we tend to ascribe higher-level skills to intelligence, including problem solving, abstract thinking and planning, language use, tool use, self-awareness, creativity, and flexibility. In addition, there seems to be a correlation between social survival and complex brains. Spindle neurons, enormous cells that allow rapid communication across the relatively large brains of great apes, elephants, and cetaceans, have been linked to intelligent behavior across species, a potential example of convergent evolution.

Within the human species we typically measure intelligence or IQ in a battery of written tests. Controversial issues are found within our own species' IQ measuring that include racial, educational level, and gender-biased questions and responses. Howard Gardner, in *Frames of Mind,* describes different types of human intelligence that include social, musical, and emotional intelligence.

There are two primary ways we measure intelligence in other species; one is physical and the other cognitive. Physically, dolphins have the second-highest brain-to-body ratio next to humans. Evolved to its current configuration for more than twenty-five million years, it differs slightly from the modern human brain. Dolphins have a large and dense corpus callosum, the structure that sends signals back and forth between hemispheres; their brain is heavily convoluted and dense with neuronal connections. Dolphins have large frontal lobes, which are thought to be the area of association and higher thinking. Dr. Lori Marino at Emory University has measured the brain-to-body ratio and next to humans dolphins have the second-highest encephalization quotient or EQ, even higher than the great apes. The idea being that dinosaurs could get along with small brains despite their large size, but anyone with a large EQ must be using all of the energetically expensive brain on something complex. Dolphin brains contain mirror neurons. These are a special class of cells in the brain that fire when an animal hears or sees an action and when the animal carries out the same action. We feel what the other is doing through these cells. Mirror neurons are thought to be involved in both language and emotional decoding and are found only in a few social species including humans, primates, and birds. The glue of complex social societies might also involve the need for emotional intelligence and mechanisms that encourage cooperation and empathy. Other sensory modalities pose interesting possibilities: could dolphins feel what other dolphins are feeling when they "hear" a whistle or a scream in the water? The only difference is modality. Such cells may play a role in the ability to share feelings, a key element of empathy.

Of course we are human, and most of what we do is from our perspective. In Temple Grandin's book *Animals in Translation,* she stresses the need, to view things from the animal's point of view. Such orientation has led to improved and less stressful conditions in even the grueling context of slaughterhouses. The point of perspective taking may be necessary for many things if we are to understand the diverse world around us, including a nonhuman-biased intelligence.

Cognitively and experimentally dolphins show advanced abilities such as abstract thinking and problem solving. Some of the most interesting work on language comprehension and abstract thinking has come out of the Kewalo Basin Marine Mammal Laboratory in Honolulu, Hawaii. Years ago, Dr. Louis Herman and Dr. Adam Pack had demonstrated the dolphin's uncanny ability to learn and comprehend both a gestural and acoustic language with their star ambassadors Phoenix and Akeakamai. Hidden behind a black screen with their white gloves poking through a hole, researchers presented the dolphin's gestural commands on an underwater TV. The dolphins only saw the motion of the white gloves contrasted against a black screen, essentially following a white bouncing ball image. The dolphins were able to extract the complex information of their commands from this motion alone. Graduate students tested with the same image failed miserably at the task.

Marc Xitco and colleagues at Disney's Epcot Center in Orlando, Florida, have tested the problem-solving abilities of bottlenose dolphins. Here the dolphins worked to solve problems involving various tasks designed to score fish from a maze of technology such as pushing levers and retrieving objects. Other researchers have demonstrated evidence of self-awareness using underwater mirror tests on bottlenose dolphins (Dr. Diana Reiss, Dr. Lori Marino, Dr. Ken Marten) and on orcas (Dr. Fabienne Delfour). Even within these human-designed and perhaps primate-biased tests, dolphins score high. Imagine if we could design an acoustic test, rather than a visual (primate-centered) test, for dolphins or other species? What would that show us? Are mirror tests the best test for an acoustic creature? Is the mirror test the best test for self-awareness? If we really want to measure a species' intelligence it might be better to develop a species-specific test based on the sensory system of the animal, such as a pheromone test for dogs or an acoustic test for dolphins.

What if an individual just isn't in the mood to take a test? Trainers in many facilities both talk about and recognize individual dolphins and their personalities; animal personality is an emerging area of study in many species including dolphins, horses, and even octopuses. In cognitive tests some individual dolphins are recognized as "ambassadors," those who are quick and willing to work with an artificial system of testing. In my own two-way work in the wild, Caroh, Tink, and Rosepetal proved to be the ambassadors of the group, adapting their schedule and their friendships to pursue another worldly interest. It is likely that most social mammals are unique individuals who may cope with and handle problems differently. So when we are measuring intelligence we might consider *whom* we are measuring, not just what species we are measuring. We must take into account the appropriateness of the test, and the individual that is tested, if we really want to unravel the minds of other species.

Many scientists believe that it is the challenge of living in a complex social group that drives the need for intelligence. One has to have long-term memory, remember individuals within the society and their habits, and communicate effectively to negotiate and resolve conflicts. The abilities needed to learn and pass on knowledge, specific knowledge unique to a local culture, could be advantageous. How might we begin to recognize other types of intelligence in nonprimate species? It is easiest to recognize types of primate intelligence primarily because given our common ancestry and evolutionary trajectory, we see similar facial expressions and body language. However, convergent evolution, the theory that species under similar selective pressures reach the same end point, leads us to think about other, nonprimate species. There is evidence of physical continuity between species, like the hydrodynamically efficient streamlined shape of sharks and dolphins. Evidence of mental continuity may be found in species faced with parallel social politics, individual personalities, and group living, such as dolphins, elephants, birds, and, of course, a variety of primates. I suspect we will also find emotional continuity in many species. Usually evolution and natural selection favors those traits that are adaptive. It stands to reason that continuity in intelligence or emotional expressions may be adaptive and something more commonly found in the natural world than we have been, up until now, willing to admit.

Studying wild dolphins to the level possible with terrestrial species is challenging. How is all this intelligence put to use in the wild? We do know that most small dolphin species have complex social structures and communication, although their "language" is yet to be deciphered. While acoustics has been a focus for most researchers, language can also be of a nonverbal nature, such as American Sign Language for humans, which encodes complex ideas and exchanges. We tend to recognize only language similar to ours, but we have to diversify our concept and analysis to really explore intelligence in other species. Cognitively, dolphins are both flexible and creative. Yet the process of enculturation (by human contact) and exposure (to other communication systems) can be contributing factors to successful experiments in captivity. Primate studies have suggested that although some primates have the abilities to do complicated tasks, it is the exposure and embedding into a human social context that encourages the emergence of intelligent actions. Could this be the case with dolphins? Do dolphins, like primates, have the structure and flexibility to learn, but only show the emergence of these features when put into the right contextual or cultural circumstances? In his sci-fi book *Speaker for the Dead,* Orson Scott Card proposes the scenario of a galaxy full of intelligent beings, some of whom are intelligent on their own, and some who have the "propensity to become intelligent with help," who are called *Varlese.* These are the species that the protagonist, Ender, is in search of, but along the way he fails to recognize the signs of potential intelligence and instead wipes out a species that were, in fact, *Varlese.*

Can we truly look at a nonspecies-biased definition of intelligence in a nonbiased, nonhumancentric way? To be intelligent, a species would probably have to be social, have a wide range of and control over sounds and signals, and use language as a subset of communication that has the ability to communicate abstract ideas. During my own career in behavioral biology, spanning thirty years, remarkable discoveries have been made illuminating the complexity and intelligence of many species. When I was in graduate school in San Francisco there was a dolphin named Spock at the marine park. When he retrieved "garbage" from his tank, such as paper and floating objects that drifted in or were thrown in by visitors, Spock was rewarded by trainers. One day a trainer noticed that Spock was bringing up rubber strips to gain his tasty fish reward. Upon observation through the underwater viewing port, Spock was seen tearing off the rubber caulking around the windows and had a stash of rubber "currency," which he would trade at the surface when he wanted a fish. Spock had learned and used the system quite well to gain an advantage. Perhaps such unexpected observations will prove the most telling in our search for what another creature's mind is like and how it operates. Equally important may be seeing spontaneous behavior, which shows us that the animal has used creativity in these challenging communication scenarios. I suspect that most nonhuman animals are both more intelligent and communicate in more complicated ways than we know.

The way animals encode specific information into their communication signals is still a mystery. From my own work trying to catalog and decipher dolphin communication I can say that it is both a difficult and long-term endeavor, but not impossible. There are some creative ways to determine the complexity of signals. Working with a set of dolphin whistles and information theory, Brenda McCowan and colleagues have calculated that the diversity and complexity of these whistles fits the criterion for "language." In another study of dolphin calves during development they also describe dolphin "babble," the equivalent of learning and honing a language over time. Such techniques are also used in various decoding algorithms to determine noise from information. Imagine if we included other types of sounds such as burst-pulsed sounds and their ultrasonic frequencies to that complexity? We might indeed find another complex and detailed language akin to the diversity found in human languages.

If other species meet the "language" criterion of intelligence, what will we do? Will we still argue that because they can't build buildings (well, beavers do), or develop mathematical formulas (or do they and just use them versus writing them down), that they still have no rights because they aren't like us? Do we have the capability to look at and act morally and ethically with species that we don't understand? Thomas White, in *In Defense of Dolphins,* argues for the rights of dolphins based on many of our own criteria for personhood. Humans, after all, are not the only species on the planet, but we often act as if we are the only species that matters, and this has severe consequences for other species, ecosystems, and the planet itself.

What are the implications of studying a nonhuman, nonprimate intelligent species? The deciphering of a nonhuman species language could be one of our greatest feats. What is more intriguing, the differences or the sameness? Will it make us more unique or bring us into a larger community of beings, of nature? The possibilities for communicating with another species abound on this planet, and dolphins top the list, despite their nonprimate appearance. In an ironic twist, the late Douglas Adams in his fourth book of *The Hitchhiker's Guide to the Galaxy, So Long, and Thanks for All the Fish,* postulated that dolphins are the more intelligent life-form, merely observing humans for their own knowledge. It may yet be our best training ground for exploring the cosmos for other life, for if we can't understand and interact with life on this planet then there is no hope for our exploration of the galaxy. That day will surely come and how we face it is up to us.

Getting close to animals in the wild involves some serious ethical issues and requires attention to the etiquette of interacting. I had thought about the consequences of even benign observation of the dolphins in my early years when I was at a lecture by Jane Goodall at the California Academy of Sciences in San Francisco in the early 1980s. After her lecture, I raised my hand and asked, "If you knew what you know now, about the impact of people and your presence with the chimpanzees, would you do it again?" I wanted to know, from this role model of mine, whether the ends did justify the means for the animals. She answered, "I think it's important that we have information about these species and someone has to be out there collecting it." So although we work in the water with the dolphins we have always tried to avoid touching, feeding, or harassing the dolphins since they already come around for purely social reasons. We work hard to recognize and respect the dolphin's social signals when we are in the water and to not encourage physical contact. A female in estrus, chased by a group of courting males, may become tactile and solicit human touch, but it is not appropriate to respond to her solicitations. When a young, playful calf interacts with us, we break off interaction if the mother appears to be trying to leave with her calf or trying to calm him down. What may appear to be the desire of a lone dolphin may not be good for the larger society. Through the years this issue has caused lively discussion aboard my boat.

In most parts of the world it has been fatal for friendly dolphins to interact with humans. Feeding wild dolphins leads to abnormal and competitive behavior between dolphins and encourages them to rely on an artificial food source. They develop trust and trust is broken. Dolphins are inevitably speared, harpooned, or wrestled by an unrespectful human being at some point. In the late 1990s there was a friendly female bottlenose dolphin in Brazil. She regularly swam along the shore and befriended swimmers. One day, two American men jumped on top of her and tried to stuff a cigarette down her blowhole. In self-defense, she fought them off, and in the process, one of the men died of internal injuries. The immediate reaction from the public was, "This dolphin is dangerous; she should be removed, relocated, so that the public is safe." Yes, that was a common reaction. I think you can see where I am going with this discussion. Shortly after the news came out a colleague made the comment, "What do you think would happen if you jumped on a large man in New York City and tried to stuff a cigarette up his nose?" Luckily, through the dedication of a local biologist, Marcos de Santos, who was well aware of the issues, no punishment occurred to the dolphin. She was allowed to stay in the area. I am unsure of whether she regained her trust or interest in humans.

It is not unnatural for us to want to interact with wildlife and the popularity of ecotourism bears witness to this desire, but even well-meaning ecotourism can often turn a beautiful field site into a circus. This has specifically been noted in the Pacific Northwest where researchers including John Ford, Alexandra Morton, and others have voluntarily removed their research boats from the area due to too much pressure on the resident killer whales. Ecotourism argues that we will save what we love and therefore we have the right to get close to nature. Successful examples of ecotourism with less of an impact include limiting numbers of hiking permits in the Grand Canyon. In my own study site it is the distance from shore that has saved the spotted dolphins from too much human impact. There are few places in the world where interspecies communication and interaction is truly a phenomenon, but the lagoons of the gray whales in Baja California is one and the sandbanks of the Bahamas where spotted dolphins live is another. These are areas that should not be overtaxed by ecotourism, but instead should be preserved and studied for their uniqueness.

To some, the alternative to seeing dolphins in the wild is to view them in captivity. Captivity takes many forms and shapes, some better and more regulated than others, but all involve the domestication of dolphins. The industry is primarily about maintaining dolphins for human purposes, including entertainment. The word "captivity" has recently been replaced by the phrase "dolphins under human care" by the industry, but the issues should not be confused or distorted. "Dolphins under human care" should apply only to dolphins that are in such care because of an injury or stranding, which requires such care and contact from either natural causes or human impacts. Captivity should apply to dolphins in confinement for human entertainment or use, whether bred for this purpose or caught from the wild. Changing the words does not change the conditions or the intention of the industry. The recent documentary *The Cove,* where fishermen from Taiji, Japan, continued their hunt to herd and slaughter hundreds of dolphins and small whales, brought to light the purchase of young female dolphins by aquariums in Asia and Eastern Europe. Primarily fed by a few American dolphin brokers, such trafficking increases the value of wild caught dolphins around the world. The same brokers traffic dolphins from the Solomon Islands to facilities in Mexico and other places as a profit-making venture. How then is the public to know what facilities encourage such practices? Will a facility tell you that the young female you are swimming with was ripped apart from her mother during capture and watched her brother die on the transport plane, a not uncommon end to international trafficking of wild animals? Although laws in the United States and Western Europe discourage the capture of wild dolphins for swim programs and human entertainment facilities, animals are sometimes housed temporarily, gaining them a "not wild" status on the books. Too often unsuspecting tourists engage in captive swim programs in Asia, the Caribbean, and other places, not realizing that most of the dolphins were captured from the wild, after witnessing the brutal slaughter of their families in places like Taiji, Japan, where young, trainable dolphins are chosen for swim programs around the world. I realize I will upset many of my colleagues who do not see this connection or a direction of action to change these horrors, but sometimes things are just wrong. Capturing wild dolphins for human entertainment is simply wrong. In many cases is it both brutal and horrific.

What possesses the good-hearted, caring people at substandard dolphin facilities to keep quiet about captures, cruelty, neglect, or death? What are the perils of a scientist speaking out against an industry that compromises the integrity of a conscious species? Ask Jane Goodall and

Roger Fouts. Ask them how they dealt with the very large industry of medical uses of chimps, a well-oiled machine deeply embedded in science, culture, and economic issues. Dr. Roger Fouts, in his incredibly powerful book *Next to Kin*, describes the pressure of the "group." Whether it is science, commercial industry, whatever the rules are, we are pressured to compromise parts of our integrity to survive in a system. In Roger's case, he describes his personal demons and struggles in an academic career, while watching the horrific conditions in which many primates are kept. His struggle led him to a personal challenge, but finally caused him to speak his mind and express his pain for the deplorable conditions.

I have seen distortion of the facts on both the industry and activist side. Although many animal advocates are trying to speak for the "voiceless," equally disturbing to me are some of the activist statements that assume that people who work in the industry are evil and unable to compromise. I was also horrified to hear, at an industry-based conference, a purposeful presentation based on "using statistics to undermine activist activities." In the same conference I heard a veterinarian representative from a major theme park talk about why it was okay to take from the wild in other countries. When I questioned him and stated my concern that, since there are not even basic population estimates from these areas, and it is irresponsible to take from them, he shrugged it off, saying, "Well, it likely doesn't have an impact." This is a total double standard and also a self-serving one. If the well-meaning, average, dolphin-loving person were exposed to the real facts of captures, deaths, and training and habituation techniques, they would be outraged.

It should not be surprising that smart animals can go crazy when confined in small and unnatural environments. In the winter of 2010 a young trainer from SeaWorld was dragged under the water and killed by a killer whale named Tilikum. Captured from the wild, Tilikum had previously been involved in three human fatalities. Tilikum is a valuable breeding male, valuable because he can create more young orcas for the industry. What happens when an intelligent animal is kept in a confined space for its long life? Isolated from their own normal whale society and its wild environment, how could a mind not go crazy? When an elephant kills a trainer I think only of the brutal treatment and beatings this smart individual has undergone most of its life. Marc Bekoff, in his book *The Emotional Lives of Animals,* and Gay Bradshaw, in her book *Elephants on the Edge,* both explore the case for mental disorders in intelligent animals including post-traumatic stress disorder (PTSD). Perhaps we should not be so surprised when an animal with a complex brain and processes parallel to our own is put in an abnormal and unhealthy circumstance and develops pathologies and disorders. Perhaps the feisty dolphin in a swim program remembers witnessing the slaughter in Taiji, Japan, of her family members while dolphin trainers grabbed her away to sell into a swim program thousands of miles away. How could you restrict a large mind in sterile and constricting environments and not have this effect? The very qualities that make us curious about these intelligent and social species are the same qualities that provide the fertile ground for mental disturbance. It would be ironic if it weren't so sad.

There is no doubt that well-meaning trainers and handlers often form strong bonds with dolphins and whales in captivity, and that this perhaps moderates the trauma of the loss of the dolphin's social group. Even though dolphins can form strong bonds with people, their lives should not be compromised for human uses such as therapy. There are many facilities that promote dolphin therapy to cure human ills including autism, depression, and more. Kristin Stewart and Lori Marino review the efficacy of therapy in a policy paper called *Dolphin-Human Interaction* and the conclusion is horrific. Hundreds of facilities have sprung up around the world, many capturing wild dolphins for this purpose, to provide what appears to be noneffective therapy for desperate parents and their children. There is no evidence that dolphin therapy is any more effective in helping humans than domesticated animal therapy, such as the interaction of dogs, cats, and horses with disabled or autistic humans. Nature is in itself healing on many levels, but dolphin therapy might even be a hindrance to children, keeping them from receiving more effective treatment with proven methods or domesticated therapy animals such as dogs or horses.

I once had a woman "reverend "call me seeking advice on starting some dolphin work. She had heard and seen some of the interactions between captive dolphins and autistic children and she wanted to start a facility to help autistic kids. "Where will you get the dolphins to start such a facility?" I inquired. "Well," she said, "we'll just take them out of the ocean." When I argued all the reasons why that isn't okay, including taking a wild dolphin away from its own family, she replied, "I think that a human child's needs are more important than the dolphin's, so I don't really have a problem with that." And thus we are faced with the essence of the issue. Animals, regardless of their needs or their levels of intelligence or family bonds, are still thought of as either property of humans or put on the planet for our use because we are human. There are some movements that lobby for the rights of great apes, our nearest relatives, but dolphins, perhaps because of the illusion of their permanently happy demeanor, have little support.

Some facilities argue that the dolphins have all their needs met, and that they are medically and physically taken care of, but still they are property, often owned and sold by mega-profit-making corporations. Some "open ocean" facilities argue that their dolphins are physically able to leave at anytime yet don't. But physical freedom is not psychological freedom and few social animals, because of their fear and psychological needs, will venture from strange but protected facilities. In many cases dolphins are captured from far-away oceans and have an abnormal social group or unfamiliar setting—we may see the illusion of physical freedom, but they are psychologically captive.

There must be nothing as devastating to a domestic animal as abandonment and betrayal by their person. The destabilization and trauma of change via disaster, death, or betrayal cannot be underestimated with an intelligent and social mind that has experienced a bond of companionship. It perhaps marks the divide between the family that has a pet as a "commodity" or one-way companionship of convenience who can be disposed of when it is inconvenient versus a family who includes a pet as part of the family and who recognizes the complexity in emotions and personality of a valued individual. It is a divide I would like to understand better, and have tried, but I sometimes think that it is the maturation of our own minds and souls that simply lead us there, and that no education in the world can change it. I hope I am wrong.

When we come to understand dolphins as sentient, social beings that require a large community and varied life experiences, where does this leave a captive dolphin? Dolphins have few options once they are in captivity. If they are stranded and able to be released they may become release candidates. This is dependent, in the United States, on their species, their age, location of stranding, and viability of feeding themselves or being released into a group. If not releasable, they are put into facilities with their own species if possible and will live out their lives either doing shows or engaged in research or therapy, depending on the facility. The reality is that most dolphins in captivity right now are not release candidates. Although some believe that releasing dolphins from captivity, no matter what the circumstances, is better than remaining in captivity, I am not one of them. There are issues around diseases, genetics, and socialization of releasing most captive dolphins that preclude this possibility. Even if the release of a dolphin back into the wild is not possible, that does not mean we should give up on releasing or retiring dolphins to a more dignified life. And I am deeply concerned about how dolphins are sold as property, treated, bred, and used for public entertainment.

So the question remains, how can we give dolphins in captivity the ability to live out their lives in dignity with their own kind in a stimulating environment? There have been multiple failed attempts at creating dolphin rehabilitation and retirement facilities, usually stemming from human ego and politics. With enough money and the right intention it is conceivable that we could create a dolphin retirement center. Models for successful retirement facilities for other social and sentient species, including primates (Center for Great Apes) and elephants (the Elephant

Sanctuary) exist and are successful alternatives to an undignified death or the cruelty of isolation. A dolphin sanctuary could relieve the burden for facilities of dolphins destined to swim alone in small tanks until they die in the back room, unnoticed and forgotten. It could also provide a place for research or education, based on the personality profiles and needs of the dolphins. Dolphin birth control could be provided if the goal of the facility was not to create more dolphins for shows, but to let these individuals live out their lives with respect. Some dolphins might choose not to interact with humans and others might step up to be ambassadors of creative work and help us understand the mind of a dolphin. Either way, it could be a win-win as long as the dolphins, not profit, remain the priority.

I do wonder what we are teaching our children when they observe captive dolphins. Are we actually modeling the wrong ethic? Even though I'm not against captivity (at least until there is a retirement center available for existing dolphins), I am against the capture of wild dolphins. By displaying dolphins for profit we teach children (and adults) that it's okay for dolphins to be our entertainment, to jump and leap on command, to tow people around in the water while we are holding a dorsal fin or, even worse, riding a dolphin or orca in a vulgar display of human dominance. We are modeling to our children that it's okay to keep animals in captivity, even large, intelligent animals with long-term memories, long-term relationships, and complex minds. Yet I'm a realist and until a retirement or rehabilitation center opens (which will be a long, expensive process) for dolphins, and as long as consumers demand entertainment and therapy from dolphins, there will be captivity in many forms.

Dolphins belong in the wild. An average spotted dolphin travels around ten to twenty miles a day, has a large network of friends and family, and invests years in teaching young dolphins how to survive. Dolphins teach their young complex skills ranging from feeding, babysitting, and negotiating the fine lines of dolphin behavior. A dolphin can only reach the true potential of a healthy, normally fully actuated individual in the wild.

Striking examples of the long-term effects of captures on wild dolphins exist. One is from the resident bottlenose dolphin community in Sarasota Bay, Florida, when such activities were allowed. Over the course of years of severe capture activity, gaps have emerged where young female dolphins are clearly missing from the reproductive pool. In a pod of killer whales in the Pacific Northwest, whales born between 1960 and 1970, almost an entire generation of orca was taken to captivity (or died in the attempt). These examples of captured individuals represent true gaps in their wild societies and because they are slow reproducing mammals, could cause their eventual demise.

It's unfortunate that the captive industry has and continues to expose people to the entertainment aspects of dolphins. To this day, most people I know have a misguided and deep-set belief that dolphins are always happy, that jumping through hoops, doing tricks, and eating dead fish is all quite normal. We come to expect this image of the performing dolphin. In the wild when people see dolphins jumping they assume it is playful, happy behavior. From my experience the dolphins are usually fighting underwater, resolving conflicts in the course of normal dolphin politics. A colleague, taking an informal poll at a Bahamian swim-with-dolphin facility, reported that the majority of tourists coming to the facility wanted the dolphins to jump more, do more aerial behavior, provide more entertainment. I can assure you that dolphins are not this way in the wild. Dolphins might allow us to watch their behavior, or interact occasionally, but they do not exist to entertain us. The saddest aspect is that we have failed to present what dolphins really are in our attempts to educate, and intentionally have manipulated the image of dolphins for profit.

Years ago, in an ironic twist, the captive industry went on an "anti-wildlife watching" campaign to advertise the harm of interacting with wild cetaceans. Their slogan was "Captivity is the place to see them because you are not harassing them in the wild." In essence there was an industrial move to make it "politically correct" to not only view dolphins in captivity but to swim with them in a tank. The fact is humans and dolphins have interacted for thousands of years. The real problems revolve around the challenges of appropriate and thoughtful etiquette and interactions with wildlife. The problem, of course, is that we cannot all expect to have the right to go anywhere or do anything we please. By encouraging viewing in captivity we take away the opportunity for people to learn "etiquette" skills in the wild, further depowering them and making them accept this view of what a dolphin is in a tank and under human control.

Besides confinement by humans and intentional captures and hunts of their families, dolphins face a variety of environmental and manmade threats. Human impacts include pollution, climate change, loss of habitat due to human development and practices (the Yangtze River dolphin just went extinct owing to the building of the Three Gorges Dam in China), and, as recently evidenced, environmental disasters including the unprecedented oil spill in the Gulf of Mexico. Apparently humans operate not on what information they know but on their feelings. Is this an example of intelligence-gone-bad or just a denial of the facts based on inconvenient changes? Most human parents will tell you that children follow what parents do, not what they say. Is the modeling of bad behavior, or unethical behavior toward other species, a potentially lethal and nonadaptive transmission of information within our species? Witness our wanton destruction of the environment, our resistance to hybrid cars and recycling in many places, the modeling to our children of our willingness to turn away as our corporate entities pollute the oceans and streams. The dangers of climate change involve not only chemical changes including endocrine disruptions, but cultural disruptions within species, and interspecies genocide. The word is not in yet as to whether the human race will survive. Humans can learn about other ways of being on this planet, from other creatures that are sentient and social, but only if we are listening.

We know chimpanzees and gorillas can comprehend and express thoughts using symbolic communication systems. Dolphins do the same. Other species have yet to be tested. I have no doubt that dolphins are intelligent, emotional, and thinking beings. Science is often the last to prove such statements because as humans we are often resistant to changing our paradigm or considering such options. Our job as scientists is not to discard phenomena on an a priori basis because of a prejudice, but to explore and study phenomena and understand the natural laws and processes that produce such phenomena. Today's view on nonhuman intelligence is yesterday's flat earth.

Spending time with wild dolphins, as individuals and as a culture, is probably as close as we can get to an understanding of the dolphin mind. I am always touched when a member of the public approaches me to thank me not only for my work but *how* I do my work. A noninvasive model, that incorporates another species freely as a participant, may be rare, but it is much needed. Why are individuals in a society important? In our own human society, an individual can move a culture, or its concepts, forward, as even other species such as Kanzi, the bonobo, Alex, the African Grey parrot, Akeakamai, the bottlenose dolphin, and Caroh, the spotted dolphin, have done for nonhuman animals. Every species has its ambassadors to the human world.

Human politics and human choices are still deeply rooted in the idea that humans are dominant over all of nature. We are a creative species whose primary talent is surviving and building environments not normally tolerated without technology. In some cases the domination of nature is supported by deeply held religious beliefs that humans have the right to do what they will with animals, plants, and even the planet. Dominance over the other is the same attitude the early explorers had when they met native cultures in other lands. We must begin to question our everyday lives and the responsibilities we have to other beings on the planet.

Paradigm shifts come hard in our culture. We attach ourselves fiercely to our realities. New ideas threaten our way of life, as did the emergence of the ideas of Galileo and others. As the evidence for nonhuman animal thinking emerges we rethink our eating habits, our treatment of pets, our willingness to encourage the captivity of at least higher mammals such as cetaceans, elephants, and the great apes. Fierce battles emerge and lines are drawn. Researchers are often pressured to explore areas that are scientifically acceptable and reject those that are believed to be unworthy of exploration. As the science, and our understanding, of consciousness and of nonhuman animal thinking proceeds and takes on rigor and interdisciplinary perspective, we will see incredible strides in our awareness of the world. Perhaps we are one of the only species that thinks of itself as separate from the natural world. It's ironic that both the health of our planet, and our individual health and well-being, may be tied profoundly to our treatment of the Earth itself. Enriching the lives of animals and sparing suffering can also increase the well-being of people in every society. Incorporating ourselves into a larger community of beings has many positive repercussions. Encroaching on an animal's habitat and polluting its environment is a reflection on the quality of our own environment.

The Future: In Their World, on Their Terms

Elizabeth Marshall Thomas, in her beautifully written book *The Tribe of Tiger*, describes the relationship between the local lions and the people in her hometown. The watering hole was a source of life to many, including the lions who came close to the houses and farms in the area. Everyone had access to the water including the lions, the people, and other species. Everyone tolerated everyone. Upon returning to this area twenty years later, Thomas noticed a change. A fence had been built around the watering hole. People had become sedentary instead of nomadic and now had permanent farms, which meant they needed, and wanted, continued and exclusive access to the water. Now, instead of allowing space for the wildlife, people were wary and showed fear of lions. They armed themselves for the approach of the lions. The lions, once calm and respectful of people, now showed fear of and aggression toward humans. A battle over resources: is there an older story? This change occurred within three generations of lions. The new generation of lions now had a different history; their own cultural context for interacting with people was based on fear and competition, unlike the previous generations.

The dolphin world, like my world, has many boundaries. I awake at 6:00 A.M. every morning, eager to catch the transition from night to day, and drink the cup of coffee that helps my brain make the transition from asleep to awake. The bright white moon set against a sapphire blue sky fades away as the sun rises. It reminds me there is no line, just continuous movement from night to day on the sandbank and all that it brings. At twilight the dolphins begin their transition, moving along the edge to feed after regrouping their forces. They meander over to the seventy-foot edge to feed and then return to the shallow sandbank. It is as if they, too, consciously assess every crossing into a boundary.

The dolphin's diurnal-nocturnal boundary is also marked by depth. Depth to a dolphin probably means safety or potential danger. To remain shallow is to see all corners, all approaches, to have maximum illumination, both from the sand and from three dimensions. Moving into the deep means being willing to perceive and experience that sense of void, knowing that something is there, but unable, or unwilling sometimes, to grasp it.

Dolphins listen a lot. They listen in clear water and in murky water. The listen as they go over the edge. If something catches their attention, they scan, investigate, and communicate. They make sense of the world around them by hearing its many subtleties. These are the subtle skills we, too, develop when spending time in nature. Some call it intuition; I call it reawakening the senses. We are our history, our evolution, and our species is built around our abilities to listen, with our ears, eyes, and hearts to the environment. Living on the edge involves the transition between boundaries, shallow and deep, light and dark, air and water. Transitions are never easy. They shake us out of our routines, out of the known world we live in. And sometimes they throw us over the edge, as we grasp for order and some sort of meaning.

Of the dolphins I first met in 1985 a handful are still around. Paint, now a grandmother three times and still reproducing herself, leads the northern group; Little Gash and Mugsy, along with Blotches, with six grandkids, lead the central group; and a few old males like Romeo and Big Gash are still around and known fathers. Romeo is also a grandfather. What is it like for a dolphin to live for fifty years and see his extended family die, get attacked by a shark, or become ill? If Romeo could tell us a story it would be a good one. I wonder if they think about how long they have known me? Has the novelty of seeing humans in their environment worn off? If so, I feel that at least the friendships remain. When I look into the eye of Paint or Big Gash, I see recognition of myself, partly expressed by a calm and collected presence. Younger dolphins in the community, although friendly, have other humans to choose from and interact with in their home. Yet we give them the space and respect we have tried to maintain with their elders over the decades, hoping that it will allow them to be comfortable.

This dolphin society is a changed community since the hurricanes of 2004 and 2005, yet their essence remains the same. Whatever knowledge was left with the surviving dolphins has been shared and transmitted to a younger generation, in hopes that it will help them in their survival in the future. Dolphins and weather willing we will continue to track a fourth generation of dolphins and watch Paint's offspring and Navel's nieces grow up in the dolphin society. We will watch which dolphin lineages continue and how their personalities develop, as we did their mothers'.

The two-way work showed me that it is possible to communicate with another species in the wild. But to create a two-way bridge between humans and another intelligent species will require dolphin-friendly technologies, yet to be developed, and other resources to allow time at sea. I plan to restart this work in the near future, dolphins and resources willing. Cracking the code and making direct strides into understanding the minds of other species could be one of our greatest accomplishments.

When I started the project I was twenty-eight years old, I'm fifty-three now, about as old as Romeo. Luckily we are both still around. I think about the life and death Romeo has seen over the decades. The loss of his friend Ridgeway, the death of many calves and other members of the community, the joys of births and the pure fun of being a dolphin on a windswept day on the shallow sandbanks. I, too, feel old, but while my species can live to around a hundred years, Romeo may be pushing it at fifty. I have had my losses, too. I lost my mother when I was twelve, my father when I was nineteen. I have lost friends to cancer and cat companions before their time. I am starting to lose friends and colleagues my age. The joys and connections we all have, as members of a community, are as real, and more similar across species, than we may want to admit. The taking of any member of a wild community has horrific implications, the loss of a mother, the interruption of a network between communities, the dissolution of a lifelong male friendship, or the loss of a sister or brother. When a human son, daughter, or father is killed in a war, not only do the family members suffer but the many friends and associates are affected. It is the same in a dolphin community. In the loss of a population of dolphins we risk the loss of a culture, something akin to genocide. Species extinction is what it says, permanent extinction off the face of the planet. I read Gay Bradshaw's book on elephants, and was appalled that 90 percent of wild elephants have been culled in what amounts to cultural interspecies genocide. The same could be said of what has happened to the pantropical spotted and spinner dolphins in the Eastern Tropical Pacific from tuna fishing—90 percent gone, millions of dolphins. Today we still intentionally hunt wolves and polar bears for fun and entertainment. Even Kruger national park, in South Africa, is "restocked" from a pool of lions and elephants kept elsewhere. It appears even our parks are bordering on the artificial. Unintentionally and neglectfully we created the largest oil spill known to us, and the consequences will be seen through the next century. We are negligent when we fail to recognize and give rights to other species, nonhuman cultures, and their societies. We all lose.

Ultimately we gain the most complete understanding of wild dolphins by watching them in their natural environment, the ocean. I wish that every dolphin might have the ability for a rich and dynamic life in the wild. Every dolphin captured for captivity, killed in a fishing net or hunt, or poisoned in an oil spill is a Romeo, a Caroh, or a Little Gash. Every dolphin is someone's mother, brother, or friend. I wish them to be always in their world, on their terms, where they belong, in the wild.

To help support Denise Herzing's ongoing research into dolphin behavior and two-way communication, go to: www.wilddolphinproject.org.

Appendix

Family Trees

Northern Family—Paint

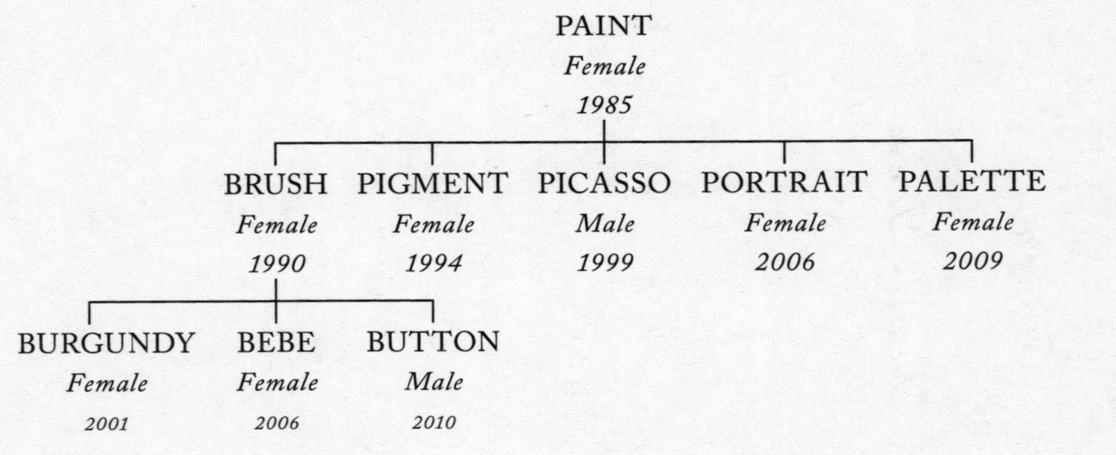

Central Family—Nippy

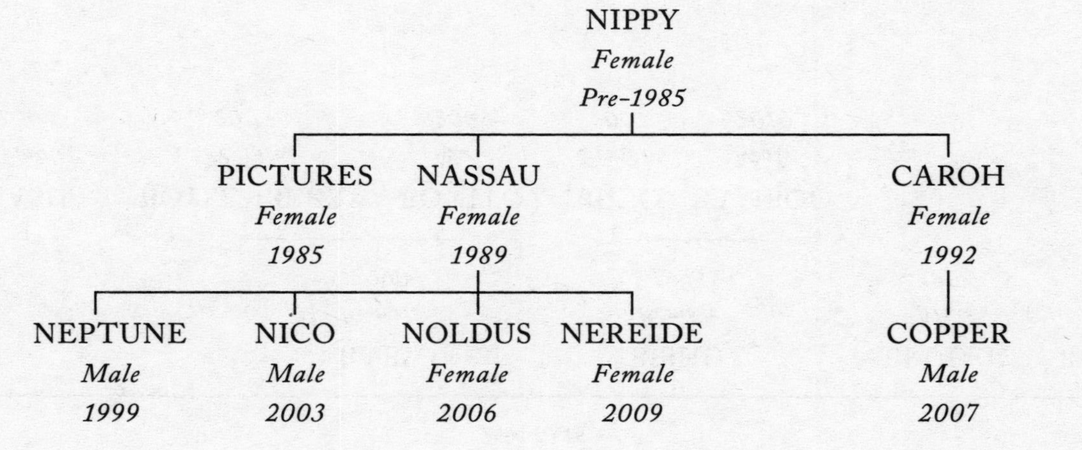

Central Family—Blotches

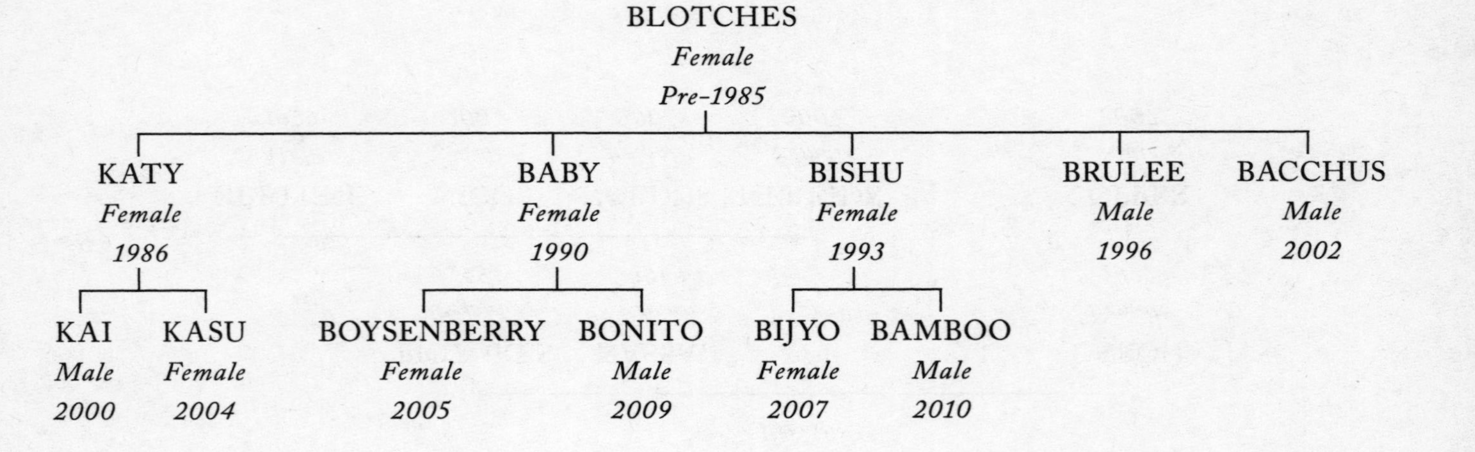

Central Family—Little Gash

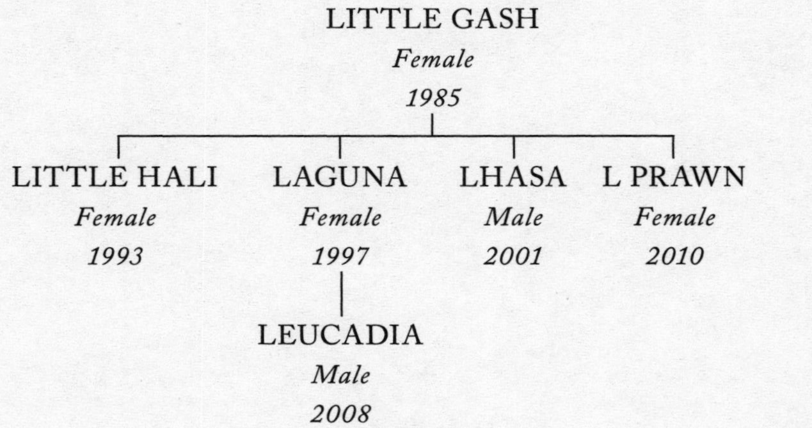

Selected References

Au, W. W. L., and D. L. Herzing. 2003. Echolocation signals of wild Atlantic spotted dolphin (*Stenella frontalis*). *Journal of the Acoustical Society of America*, 113 (1): 598–604.

Bateson, G. 1972. *Steps to an Ecology of Mind: Collected Essays in Anthropology, Psychiatry, Evolution, and Epistemology*. San Francisco, Calif.: Chandler Publishers.

Bekoff, M. 2000. *The Smile of a Dolphin: Remarkable Accounts of Animal Emotions*. New York: Discovery Books.

______. 2007. *The Emotional Lives of Animals: A Leading Scientist Explores Animal Joy, Sorrow, and Empathy—and Why They Matter*. Novato, Calif.: New World Library.

Bender, C. E., Herzing, D. L., and D. F. Bjorklund. Evidence of Teaching in Atlantic Spotted Dolphins (*Stenella frontalis*) by Mother Dolphins Foraging in the Presence of their Calves. *Animal Cognition*, 12 (1): 43–53.

Bradshaw, G. A. 2009. *Elephants on the Edge: What Animals Teach Us about Humanity*. New Haven, Conn.: Yale University Press.

Connor, R. C., Smolker, R. A., and A. F. Richards. 1992. Two levels of alliance formation among male bottlenose dolphins (*Tursiops sp.*). *Proceedings of the National Academy of Sciences*, 89: 987–90.

Damasio, A. 1994. *Descartes' Error: Emotion, Reason, and the Human Brain*. New York: G. P. Putnam's Sons.

Darwin, C. 1998. *The Expression of Emotions in Man and Animals*, Third Edition. New York: Oxford University Press.

Davis, H., and D. Balfour. 1992. *The Inevitable Bond: Examining Scientist-Animal Interactions.* New York: Cambridge University Press.

Deecke, V., and V. Janik. 2006. Automated categorization of bioacoustic signals: Avoiding perpetual pitfalls. *Journal of the Acoustical Society of America,* 119 (1): 645–53.

de Waal, F. 2009. *The Age of Empathy: Nature's Lessons for a Kinder Society.* New York: Three Rivers Press.

Ekman, Paul. 2003. *Emotions Revealed, Second Edition: Recognizing Faces and Feelings to Improve Communication and Emotional Life.* New York: Henry Holt and Company.

Fossey, Dian. 1983. *Gorillas in the Mist.* Boston: Houghton Mifflin Company.

Fouts, R. 1997. *Next of Kin: What Chimpanzees Have Taught Me about Who We Are.* New York: William Morrow and Company.

Frantzis, A., and D. L. Herzing. 2002. Mixed-species associations of striped dolphins (*Stenella coeruleoalba*), common dolphins (*Delphinus delphis*) and Risso's dolphins (*Grampus griseaus*), in the Gulf of Corinth (Greece, Mediterranean Sea). *Aquatic Mammals,* 23 (2): 85–99.

Fripp, D., et al. 2005. Bottlenose dolphin (*Tursiops truncatus*) calves appear to model their signature whistles on the signature whistles of community members. *Animal Cognition*, 8 (1): 17–26.

Gardner, H. 1983. *Frames of Mind: The Theory of Multiple Intelligences.* New York: Basic Books.

Goodall, Jane. 1971. *In the Shadow of Man.* New York: Houghton Mifflin Company.

Grandin, T. 2005. *Animals in Translation: Using the Mysteries of Autism to Decode Animal Behavior.* New York: Scribner.

Green, M. L., et al. 2007. Noninvasive methodology for the sampling and extraction of DNA from free-ranging Atlantic spotted dolphins (*Stenella frontalis*). *Molecular Ecology Notes*, 7: 1287–92.

Griffin, D. R. 1981. *The Question of Animal Awareness: Evolutionary Continuity of Mental Experience.* Los Altos, Calif.: William Kaufman, Inc.

Herzing, D. L. 1996. Vocalizations and associated underwater behavior of free-ranging Atlantic spotted dolphins, *Stenella frontalis,* and bottlenose dolphins, *Tursiopstruncatus. Aquatic Mammals,* 22 (2): 61–79.

———. 1997. The natural history of free-ranging Atlantic spotted dolphins

(*Stenella frontalis*): Age classes, color phases, and female reproduction. *Marine Mammal Science*, 13 (4): 576–95.

______. 2000. Acoustics and Social Behavior of Wild Dolphins: Implications for a Sound Society. In *Hearing in Whales, Springer-Verlag Handbook of Auditory Research*, 225–72. New York: Springer-Verlag.

______. 2004. Social and Non-Social Uses of Echolocation in Free-Ranging *Stenella frontalis* and *Tursiopstruncatus*. In *Advances in the Study of Echolocation in Bats and Dolphins*, 404–10. New York: Springer-Verlag.

______. 2005. Transmission mechanisms of social learning in dolphins: Underwater observations of free-ranging dolphins in the Bahamas. *Autour de L'Ethologieet de la Cognition Animale*, 185–193. Lyon: Presses Universitaires de Lyon.

______. 2006. The Currency of Cognition: Assessing Tools, Techniques, and Media for Complex Behavioral Analysis. *Aquatic Mammals*, 32 (4): 544–53.

Herzing, D. L., and M. dos Santos. 2004. Functional Aspects of Echolocation in Dolphins. In *Advances in the Study of Echolocation on Bats and Dolphins*, 386–93. New York: Springer-Verlag.

Herzing, D. L., and C. M. Johnson. 1997. Interspecific interactions between Atlantic spotted dolphins (*Stenella fronialis*) and bottlenose dolphins (*Tursiops truncatus*) in the Bahamas, 1985–1995. *Aquatic Mammals*, 23 (2): 95–99.

______. 2006. Conclusions and Possibilities of New Frameworks and Techniques for Research on Marine Mammal Cognition. *Aquatic Mammals*, 32 (4): 554–57.

Herzing, D. L., and T. J. White. 1999. Dolphins and the Question of Personhood. Special Issue: *Etica Animali*, 9/98: 64–84.

Horner, J. 1996. *The Abstract Wild*. Tucson, Ariz.: University of Arizona Press.

Johnson, C. M., and D. L. Herzing. 2006. Primate, Cetacean, and Pinniped Cognition Compared: An Introduction. *Aquatic Mammals*, 32 (4): 409–12.

King, B. J. 2004. *The Dynamic Dance: Nonvocal Communication in African Great Apes*. Cambridge, Mass.: Harvard University Press.

Kuhn, T. 1996. *The Structure of Scientific Revolutions*. Chicago, Ill.: University of Chicago Press.

Lammers, M. O., Au, W. W. L., and D. L. Herzing. 2003. The broadband social acoustic signaling behavior of spinner and spotted dolphins. *Journal of the Acoustical Society of America*, 114 (3): 1629–39.

LeDuc, R. G., et al. 1999. Phylogenetic relationships among the Delphinid cetaceans based on full cytochrome B sequences. *Marine Mammal Science,* 15 (3): 619–48.

Lilly, John C. 1978. *Communication between Man and Dolphin: The Possibilities of Talking with Other Species.* New York: Crown Publishers.

Lockyer, C. 1990. Review of incidents involving wild, sociable dolphin, worldwide. In S. Leatherwood and R. R. Reeves, eds., *The Bottlenose Dolphin,* 337–54. San Diego, Calif.: Academic Press.

Lusseau, D., and M. E. J. Newman. 2004. Identifying the role that animals play in their social networks. *Proceedings of the Royal Society,* 271 (6): S477–81.

McConnell, P. B. 1990. Acoustic structure and receiver response in domestic dogs, *Canisfamiliaris. Animal Behavior,* 39: 887–904.

McGowan, B., Hansler, S. F., and L. R. Doyle. 2002. Using information theory to assess the diversity, complexity, and development of communicative repertoires. *Journal of Comparative Psychology,* 116 (2): 166–72.

Malson, L. 1972. *Wolf Children and the Problem of Human Nature.* New York: Monthly Review Press.

Marino, L. 1997. Relationship between gestational length, encephalization, and body weight on odontocetes. *Marine Mammal Science,* 13 (1): 133–38.

Marten, K., and S. Psarakos. 1995. Using Self-View Television to Distinguish between Self-Examination and Social Behavior in the Bottlenose Dolphin (*Tursiops truncatus*). *Consciousness and Cognition,* 4 (2): 205–24.

Marten, K., et al. 2001. The acoustic predation hypothesis: Linking underwater observations and recordings during odontocete predation and observing the effects of loud impulsive sounds on fish. *Aquatic Mammals,* 27 (1): 56–66.

Mauck, B., et al. 2005. How a Harbor Seal Sees the Night Sky. *Marine Mammal Science,* 20 (4): 688–708.

Morton, A. 2002. *Listening to Whales: What the Orcas Have Taught Us.* New York: Ballantine Books.

Morton, E. S. 1997. On the recurrence and significance of motivation—structural rules in some bird and mammal sounds. *American Naturalist,* 111: 855–69.

Moss, C. 1989. *Elephant Memories: Thirteen Years in the Life of an Elephant Family.* New York: Fawcett Columbine.

Norris, K. S. 1992. *Dolphin Days: The Life and Times of the Spinner Dolphin.* New York: W. W. Norton and Company.

Pack, A. A., and L. M. Herman. 1995. Sensory integration in the bottlenose dolphin: Immediate recognition of complex shapes across the senses of echolocation and vision. *Journal of the Acoustical Society of America*, 98 (2): 722–33.

Parsons, K. M., et al. 2006. Population genetic structure of coastal bottlenose dolphins (*Turisops truncatus*) in the northern Bahamas. *Marine Mammal Science*, 22 (2): 276–98.

Patterson, F., and E. Linden. 1981. *The Education of Koko*. New York: Holt Rinehart and Winston.

Pepperberg, I. 2006. Cognitive and communicative abilities of gray parrots. *Applied Animal Behavior*, 100 (1): 77–86.

Perrin, W. F. Color pattern of the eastern Pacific spotted porpoise (*Stenella graffmai* Lonnberg (Cetecea, Delphinidae). *Zoologica*, 54: 135–49.

Peterson, D., and J. Goodall. 1993. *Visions of Caliban: On Chimpanzees and People*. Boston: Houghton Mifflin.

Pryor, K., et al. 1990. A dolphin-human fishery cooperative in Brazil. *Marine Mammal Science*, 6 (1): 77–82.

Psarakos, S., Herzing, D. L., and K. Marten. 2003. Mixed-species associations between Pantropical spotted dolphins (*Stenella attenuata*) and Hawaiian spinner dolphins (*Stenella longirostris*) of Oahu, Hawaii. *Aquatic Mammals*, 29 (3): 390–95.

Reiss, D., and L. Marino. 2001. Mirror self-recognition in the bottlenose dolphin: A case of cognitive convergence. *Proceedings of the National Academy of Sciences*, 98: 5937–42.

Rendell, L. E., and H. Whitehead. 2001. Culture in whales and dolphins. *Behavior and Brain Sciences*, 24: 309–82.

Rogers, C. A., et al. 2004. The Social Structure of Bottlenose Dolphins, *Tursiops truncatus*, in the Bahamas. *Marine Mammal Science*, 20 (4): 688–708.

Rossbach, K. A., and D. L. Herzing. 1997. Underwater observations of Benthic-feeding bottlenose dolphins (*Tursiops truncatus*) near Grand Bahama Island, Bahamas. *Marine Mammal Science*, 13 (3): 498–504.

______. 1999. Inshore and Offshore Bottlenose Dolphin (*Tursiops truncatus*) Communities Distinguished by Association Patterns, Near Grand Bahama Island, Bahamas. *Canadian Journal of Zoology*, 77: 581–92.

Savage-Rumbaugh, S., and R. Lewin. 1996. *Kanzi: The Ape at the Brink of the Human Mind*. New York: John Wiley and Sons.

Seyfarth, R. M., et al. 1980. Vervet monkey alarm calls: Semantic communication in a free-ranging primate. *Animal Behavior,* 28: 1070–94.

Slobodchikoff, C. N., Perla, B. S., and J. L. Verdoli. 2009. *Prairie Dogs: Communication and Community in an Animal Society.* Cambridge, Mass.: Harvard University Press.

Stewart, K. L., and L. Marino. 2009. Dolphin-Human Interaction Programs: Policies, Problems and Alternatives. Ann Arbor, Mich.: Animals and Society Institute.

Strier, K. B. 2002. Primates Past and Present. In *Primate Behavioral Ecology,* Second Edition, 73–98. Boston: Allyn and Bacon.

Sylvestre, J. P., and S. Tasaka. 1985. On the intergeneric hybrids in cetaceans. *Aquatic Mammals,* 11: 101–108.

Thomas, E. M. 2001. *The Tribe of Tiger: Cats and Their Culture.* New York: Pocket Books.

White, Thomas I. 2007. *In Defense of Dolphins: The New Moral Frontier.* Malden, Mass.: Blackwell Publishing.

Willis, P., et al. 2004. Natural hybridization between Dall's porpoises (*Phocoenoidesalli*) and harbour porpoises (*Phocoenaphocoena*). *Canadian Journal of Zoology,* 82 (5): 828–34.

Wolters, S., and K. Zuberbühler. 2003. Mixed-species associations of Diana and Campbell's monkeys: The costs and benefits of a forest phenomenon. *Behaviour,* 140 (3): 371–85.

Xitco, M., et al. 2001. Spontaneous pointing by bottlenose dolphins (*Tursiops truncatus*). *Animal Cognition,* 4 (2): 115–23.

Index